Springer Tracts in Human-Centered Computing

The book series is aimed at providing an exchange platform for researchers to summarize the latest research and developments related to human-centered computing. Human-centered computing is focused on the study of design, development, and deployment of human-computer systems based on mixed-initiatives. It may be visualized as a three-dimensional space comprising human, computer, and environment. This upcoming computing paradigm has emerged from the convergence of multiple disciplines concerned both with understanding human beings and with the design of computational artifacts. The field of human-centered computing is a multidisciplinary field encompassing disciplines such as computer science, human factors, sociology, psychology, cognitive science, anthropology, communication studies, graphic design and industrial design.

The book series covers the topics and fields of distributed environments entailing Internet-based information systems, grids, sensor-based information networks, and mobile and wearable information appliances, multimedia and multi-modal human-computer interfaces, design of intelligent interfaces and information representation and visualization, multi-agent systems, effective and constrained computer-mediated human-human interaction, defining relevant semantic structures for multimedia information representation, community specific HCI solutions, collaborative and interactive systems, social interaction, social orthotics, affective computing, knowledge-driven human-computer interaction, human-centered semantic formulation, human-centered management science, and participatory action research.

The series will include monographs, edited volumes, and selected proceedings.

Mario Köppen · Masato Uchida · Masato Tsuru ·
Włodzimierz Ogryczak · Siddhartha Bhattacharyya

Resource-Aware Intelligent Communication Systems

Springer

Mario Köppen
Department of Computer Science
and Networks
Kyushu Institute of Technology
Fukuoka, Japan

Masato Tsuru
Department of Computer Science
and Networks
Kyushu Institute of Technology
Fukuoka, Japan

Siddhartha Bhattacharyya
VSB Technical University of Ostrava
Ostrava, Czech Republic

Algebra Bernays University
Zagreb, Croatia

Masato Uchida
Department of Computer Science
and Engineering
Waseda University
Tokyo, Japan

Włodzimierz Ogryczak
Institute of Control and Computation
Engineering
Warsaw University of Technology
Warsaw, Poland

ISSN 2662-6926 ISSN 2662-6934 (electronic)
Springer Tracts in Human-Centered Computing
ISBN 978-3-032-06914-6 ISBN 978-3-032-06915-3 (eBook)
https://doi.org/10.1007/978-3-032-06915-3

This Springer imprint is published by the registered company Springer Nature Switzerland AG
The registered company address is: Gewerbestrasse 11, 6330 Cham, Switzerland

Preface

The recent prevalence of modern, high-performance communication technology, invading virtually all aspects of our social, cultural, creative, political, and economic life, stays in a sharp imbalance in how we perceive that technology as being based on resource utilization. For short, the resources needed to fuel that development appear to be unlimited, auto-replenishing, and always immediately available, services at your fingertip. But it only needs a bit of scratching on the surface of things to see that this is truly not the case.

In spite of constant efforts invested in this direction including the well-known Arrows theorem stating that any fair voting is a dictatorship, with many generalizations, leaving the impression that such studies are ill-posed and at least theoretical we will not be able to tackle a resource-sharing problem in a perfect or tractable way. The advance in computer hardware architectures starting in the 80s, and 90s of the last century fueled the basic problem of judicious sharing of CPU resources time and memory among concurrent tasks. Although a big boost to solve this issue came up with the spread of modern wireless communication technologies including ways and methods to share bandwidth, quickly respond to changing demands, service a scalable number of users, as well as provide and smartly use energy, yet several sidetracks are still not followed up like open ends, things open to other ways of interpretation and models. To be precise, the research situation is not well-centered and the attention paid is not sufficient.

This volume highlights the foundations, methodology, and real-world situations for intelligent communication systems under the aspect of resource utilization, sharing, and maintenance.

The design, configuration, and operation of technical system often leaves us with a plethora of choices for its variables. Optimality appears to be a convenient way to path a way into the configurational jungle and come up with a single solicited setting for all our needs. It misses the point that optimality is not specified per se. Chapter 1 frames this issue of design, configuration, and operation of such technical systems as an optimization problem with reference to the optimal utilization of the undelying resources to attain user satisfaction.

Chapter 2 generalizes the introduced framework in terms of a relational framework with application to Pareto dominance relations. This chapter also delves into other ideas of optimality and best choices from a choice set through the rationalization of choices.

With the advent of high end computational devices and storage systems to cater the wide variety of data capturing and acquisition devices, the amount of data being generated and needed to be processed has increased manifold calling for greater computational resources to undertake the processing tasks. This has not only led to a huge demand for resources, but also increased energy usage. Thus, it calls for resource-aware systems which can accommodate optimal resource usage under user-specified constraints. Chapter 3 illustrates several case studies for resource-aware intelligent communication systems highlighting their essence and applications.

Fukuoka, Japan — Mario Köppen
Tokyo, Japan — Masato Uchida
Fukuoka, Japan — Masato Tsuru
Warsaw, Poland — Włodzimierz Ogryczak
Ostrava, Czech Republic — Siddhartha Bhattacharyya
Zagreb, Croatia
March 2025

Contents

Symbols

R	Binary relation
X	General domain of a relation, often $X \subseteq R_n^+$
$X \times X$	Direct product of X with itself
x, y, z	General elements of the domain X
a, b, c	Symbolic elements of the domain X
s, t, u	Specific elements of the domain X
i, j, k	Indizes for ordered sets or vector components
$R \downarrow_S$	Restriction of relation R to a subdomain $S \subseteq X$
$R \uparrow_Y$	Extension of relation R to a superdomain $Y \supseteq X$
O_X	Empty relation $\emptyset$
U_X	Universal relation $X \times X$
D_X	Diagonal relation of all pairs (x, x)
$\odot_x$	Point relation of the only pair (x, x)
$\min_{x \neq y}$	Minimum value of a component $x_i \neq y_i$
$x_{[i]}$	i-th smallest component of x
2^X	Powerset of X

Chapter 1
The Basic Idea of Resource-Awareness

The design, configuration, and operation of a technical system often leave us with a plethora of choices for its variables. Optimality appears to be a convenient way to pave a way into the configurational jungle and come up with a single solicited setting for all our needs. It misses the point that optimality is not specified per se.

We consider a rather simple problem where optimality seems to provide an easy-going solution. This problem is a variation of problems studied in [6, Chap. 6]. Consider the network shown in Fig. 1.1. It is an undirected graph of ten nodes, which are connected by a number of links. Three users want to route traffic through the network along the paths indicated in the figure in order to transfer some amount of data at admitted rates. It is assumed that all communication means, like sessions, have been established, and we can focus our attention on the so-called "end-to-end" traffic rate allocation problem. The crux of the problem is that the links have a limited capacity, while at the same time, the topology of the network forces users to share links.

Of course, this way many technical details are neglected. But the example already represents the essential aspect of the related choice problem such that we can gain some knowledge of its general aspects, even if we populate it with more and more technical desiderata (or, as Axelrod writes in a treatise to motivate the study of simplifying abstractions [3, p. 19]: "It is the very complexity of reality which makes the analysis of an abstract interaction so helpful as an aide to understanding").

As long as the traffic demands of the users are not so high, traffic can flow through it without any concern. But as it can be expected, the demands will grow and soon the link capacities will become a striving issue that limits the availability of the network resource for its users. At this point, a clear cut has to be made for deciding what can be admitted to the users and what not. So, we want to assign traffic rates that are in some sense the BEST way to employ the resources and achieve what might be called "user satisfaction." We have an optimization problem.

Consider the total of the traffic rates assigned to the user. If we call it simply x_1, x_2 and x_3 then we assume at first that these values are non-negative, and that the

M. Köppen et al., *Resource-aware Intelligent Communication Systems*, Springer Tracts in Human-Centered Computing, https://doi.org/10.1007/978-3-032-06915-3_1

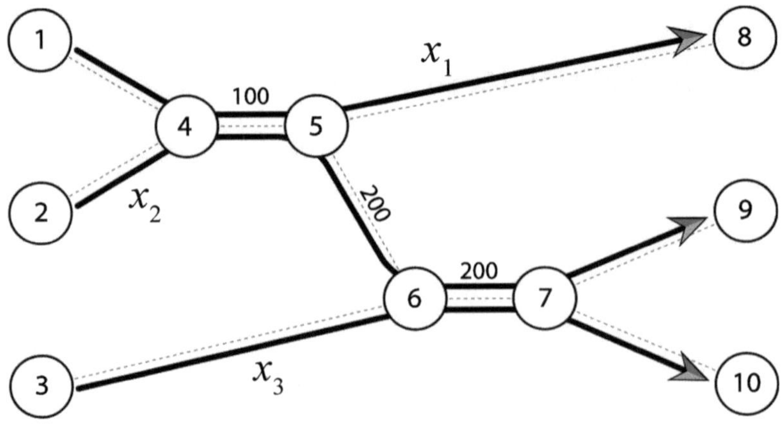

Fig. 1.1 Network routing example for three end-to-end user traffics

capacity limits apply. Another look at Fig. 1.1 shows that we have to focus on the following inequalities.

$$x_1 + x_2 \leq 100 \tag{1.1}$$

$$x_2 + x_3 \leq 200 \tag{1.2}$$

This reflects the capacity limits of the link between nodes 2 and 3, which is shared by users 2 and 3, and for the link between nodes 5 and 6, shared by users 2 and 3. So, for example, we can ignore the limit capacity of 100 of the link between nodes 3 and 5: only route 2 traverses this link, and from the non-negativity condition above, inequalities already guarantee that $x_2 \leq 100$. For reasons that will be clarified later, these two links are considered the "true" bottlenecks of the rate assignment problem.

Assume that the triple $M^* = (x_1^*, x_2^*, x_3^*)$ of traffic rates is maximizing the total rate, i.e., their sum. There, we also assume that x_2^* has a value $\delta > 0$. Now we reduce x_2^* by $\delta/2$. Since the triple $M*$ fulfilled both inequalities (1.1) and (1.2) it must be that

$$(x_1^* + \delta/2) + (x_2^* - \delta/2) \leq 100 \tag{1.3}$$

$$(x_2^* - \delta/2) + (x_3^* + \delta/2) \leq 200 \tag{1.4}$$

which means that also the triple $M = (x_1^* + \delta/2, x_2^* - \delta/2, x_3^* + \delta 2)$ is feasible. But its total is $x_1^* + x_2^* + x_3^* + \delta/2$ and thus larger than $x_1^* + x_2^* + x_3^*$—there is a conflict with assuming that the rate for user 2 is positive in the maximal assignment. The only conclusion is that $x_2^* = 0$.

Now, by adding both inequalities in Eqs. (1.1) and (1.2), we can easily find an upper bound for the total rate as

$$x_1 + x_2 + x_3 \leq x_1 + 2x_2 + x_3 \leq \tag{1.5}$$
$$\leq (x_1 + x_2) + (x_2 + x_3) \leq 100 + 200 = 300 \tag{1.6}$$

and by setting $x_1^* = 100$, $x_2^* = 0$, $x_3^* = 200$ we yield this upper bound and thus have a solution for the maximal rate.

So instead of an optimal solution, we get a result where it appears "best" to set x_2 to 0, so to shut off user 2 from the traffic routing. Even if we neglect the fact that user 2 might oppose this solution for understandable reasons, it is also hard to assume that a sum will be optimized by setting one of the summands to zero.

The underlying reason is, of course, that user 2 interferes with two other users independently. The example in [6] is basically an extension of this scenario to n users, where one and only one user shares a link with all other $(n-1)$ users. The result is the same. In the deepening part of this chapter, we will see that there are situations where one third of all rate assignments are best set to 0 in order to gain a maximum total.

Then, what goes wrong here?

If we cannot combine the assigned rates by their simple sum, we may send each user on a race by themselves. Consider an assignment (10, 20, 50). It is feasible. But we can increase the rates for all users by, e.g., taking now the rates (20, 30, 60)—where all users gain 10 units. Since there are limits on the capacity, for sure, this game cannot be continued endlessly. So we might ask for the frontiers of this race, i.e., when there are rate assignments where not all rates can grow together anymore.

In economics, this is commonly called the efficient solution. Often it is also called Pareto efficiency. Vilfredo Pareto was a late nineteenth-century/beginning twentieth-century Italian economist who brought many formal ideas and concepts into the economic sciences that are still in use today. Among them is some idea about the stability of an economy. These days, such ideas are put into a more liberal and social context, when it is said that Pareto efficiency means that one can only become better off if someone else becomes worse off. Appreciating the fact that we live today in social and economic contexts more and more characterized as "winner-takes-all societies" [9], one might ask why the one who gets better off should care about anyone getting worse off at all. In fact, Pareto considered the perseverance of elites, struggling in competition with other upcoming elites. As soon as a state is reached where the competing elite can only become stronger if the ruling elite is weakened, it gives the desiderata for the ruling elite to maintain its power.

This might be a strong way of thinking about our small three-user-in-a-network problem, but nevertheless, formally it appears like this: find an assignment such that the only way to further increase the rate for one user is to reduce the rate of another user.

Assume that in such an assignment, user 1 receives a rate of x. Then, if x and x_2 together would be less than 100, it would not be an efficient assignment. Why? Because we could still increase all rates by a non-negative amount such that at least one increase is positive. If $x + x_2 = 100 - \delta$ then this would be the case for the assignment $(x + \delta, x_2, x_3)$, which is still feasible. Thus, for a (Pareto-)efficient

Table 1.1 List of efficient assignments

No.	x_1	x_2	x_3
1	0	100	100
2	20	80	120
3	30	70	130
4	$33.\bar{3}$	$66.\bar{6}$	$133.\bar{3}$
5	50	50	150
6	57.735	42.265	157.735
7	80	20	180
8	100	0	200

assignment, it must be $x + x_2 = 100$ and then we find for the same reason that $x_2 + x_3 = 200$ or $(100 - x) + x_3 = 200$ or $x_3 = 100 + x$. Altogether we yield an efficient assignment for each x from $[0, 100]$ and only for them as

$$x_1 = x \tag{1.7}$$

$$x_2 = 100 - x \tag{1.8}$$

$$x_3 = 100 + x \tag{1.9}$$

Table 1.1 lists a number of such efficient assignments.

The entry number 8 is already known to us: it is the assignment with a maximum total of 300. The meaning of other numbered entries will be unrevealed in the following. For now, we confirm an inconvenient circumstance of the efficiency concept: it offers an infinite set of solutions. It is not so often noted that by employing optimality, in addition to the fact that a selection procedure for a setting is envisaged, it is also of some importance that the selection has a single target only. At the end, we can only use one specific setting at a time—with the probable extension that among a set of solutions we might use them in a sequence. But in no case we can handle an infinite number of solutions.

This might give the impression that the set of Pareto efficient solutions to our problem is of no value. But we will see that these solutions give a suitable frame for approaching more specific solutions. For now, we keep this solution set in mind and look for a different approach.

The main problem with the "optimal" $x_2 = 0$ assignment in the first solution is that of the imbalanced distribution of assignments. Thus, we can try to focus on solutions where the balance between individual user assignments is preserved. In other words, we want to somehow favor solutions with a low spread of their values, being as equal as possible. Actually, there are many ways to come up with a numerical representation of equalness.

One very popular approach is due to Jain [16]. It is based on the following reasoning about sets of equal values. There is a famous inequality among power means. Given a set of n non-negative numbers $x_1, x_2, \ldots, x_n$, the pth power mean is defined as

$$M_p(x_1, x_2, \ldots, x_n) = \left[\frac{1}{n} \sum_{i=1}^{n} x_i^p \right]^{\frac{1}{p}} \tag{1.10}$$

for $p \neq 0$ and for $p = 0$ it is the so-called geometric mean given by

$$M_0(x_1, x_2, \ldots, x_n) = \sqrt[n]{\prod_{i=1}^{n} x_i} \tag{1.11}$$

In general, for $q < p$ it is $M_q(x_1, \ldots, x_n) \leq M_p(x_1, \ldots, x_n)$ and the equality holds if and only if $x_1 = x_2 = \ldots = x_n$.

So if we divide a qth power mean by a pth power mean, we will get a non-negative value that is at most 1, and it is 1 if and only if all x_i are equal. Thus, such ratios serve as a good indicator for equality—or low spread—of a set of non-negative values. In the approach of Jain, the ratio of arithmetic ($p = 1$) and harmonic mean $p = 2$ was used.

$$\frac{1}{n} \sum_{i=1}^{n} x_i \leq \sqrt{\frac{1}{n} \sum_{i=1}^{n} x_i^2}$$

$$\frac{1}{n^2} \left(\sum_{i=1}^{n} x_i \right)^2 \leq \frac{1}{n} \sum_{i=1}^{n} x_i^2$$

$$\frac{1}{n} \frac{\left(\sum_{i=1}^{n} x_i\right)^2}{\sum_{i=1}^{n} x_i^2} \leq 1$$

The operator at the left-hand side of the last inequality is called Jain's Fairness Index.

Definition 1.1 Given a set of n non-negative real numbers x_i. Then the Jain's Fairness Index of these values is defined as

$$\mathcal{J}(x_1, x_2, \ldots, x_n) = \frac{1}{n} \frac{\left(\sum_{i=1}^{n} x_i\right)^2}{\sum_{i=1}^{n} x_i^2} \tag{1.12}$$

It is a value from [0, 1] and yields the value 1 if and only if all x_i are equal. It has another property that is only valid if we start from the power means for $p = 1$ and $p = 2$: assume that only a fraction f of assignments are equal to each other and all other assignments are 0. Then the Jain's Fairness Index of these values equals f. It can be easily seen: if $x_i = x$ for $i = 1, \ldots, m \leq n$ and $x_i = 0$ otherwise, we have

$$\mathcal{J} = \frac{1}{n} \frac{\left(\sum_{i=1}^{n} x_i\right)^2}{\sum_{i=1}^{n} x_i^2} = \frac{1}{n} \frac{(m \cdot x)^2}{m \cdot x^2} = \frac{m}{n} = f$$

So a Jain's Fairness Index of 0.8 can be understood as resembling a situation where 80% of the users receive the same share and 20% nothing.

If we want to apply this measure to the traffic rate assignment problem, there is one point that needs some clarification: the Jain's Fairness Index is maximal (of value 1) once all its arguments are equal. So an assignment of rates 0.01 to all three users would yield a maximal index value of 1. Obviously, this measure does not take efficiency into account. This obvious problem did not affect the popularity of this index in experimental parts of publications on network control (see, just as a random pick the usage of this index in papers like [14, 17, 39]). In [22] it is assumed that the problem can be solved by multiplying it with an efficiency-related term, thus losing the boundedness of the index between 0 and 1. A more elegant way is just to select among the efficient solutions the one with the largest Jain's Fairness Index, thus, using the settings from Eqs. (1.1) and (1.2).

$$
\begin{aligned}
\mathcal{J}(x_1, x_2, x_3) &= \frac{(x + (100 - x) + (100 + x))^2}{9 \cdot (x^2 + (100 - x)^2 + (100 + x)^2)} \\
&= \frac{(x + 200)^2}{9x^2 + 60,000}
\end{aligned}
$$

We skip the details and take the result for setting the first derivative to 0, which happens for $x = 100/3$, giving $\mathcal{J} = 7/9 = 0.\bar{7}$. To make sure that this is a maximum, we compare with the range border values: for $x = 0$ it is $\mathcal{J} = 2/3 = 0.\bar{6}$ and for $x = 100$ it is $\mathcal{J} = 0.6$. Thus, entry number 4 in Table 1.1 is the efficient assignment with the largest Jain's Fairness Index.

Referring to the former argument, this corresponds to a situation where 100% of the total wealth (where the commodities are the assigned traffic rates) is in the hands of about 78% of the users, while 22% have nothing. However, we do not need such a model when the original distribution of wealth is directly available. This leads to the concept of the Lorenz curve.

In the above solution, the lowest assignment of about 33.3 goes to user 1. So user 1 is the poorest in our small three-user community, followed by user 2 with about 67, and user 3 is the richest with an assigned rate of 133. So, we can also say that a third of the users own 133/233 or 4/7, which is about 57% of the total wealth (233 is the sum of all three rates in this assignment, 133 is the assignment received by user 3 alone). Two-thirds of all users own $(67 + 133)/233$ or about 85%.

Imagine the construction of a curve: the x-axis is the users sorted by non-decreasing wealth (like user 1, user 2, user 3). On the y-axis, we graph the so-called ordered sums: point 0 is 0, point 1 is the smallest rate. The next point 2 is the sum of the smallest and second-smallest rates. Point 3 is the total of all rates (which can also be seen as the sum of the smallest, second-smallest, and third-smallest rates). This is the so-called Lorenz curve for the distribution of rates. If we were to distribute the same total such that all users would receive the same rate, the graph would be a straight line from (0, 0) to (3, 233) as indicated by the dotted line. So a measure for deviation from equality is given by the area enclosed by the Lorenz curve of our wealth distribution and the dotted line of equal income. This area is divided by the total area of the triangle bounded by the coordinate axes and the dotted line for normalization purposes. This ratio is called the Gini index of the distribution. A smaller value indicates closer values, and it is 0 in case all values are equal.

There is a more convenient way to compute the Gini index from given values x_i (with $i = 1, \ldots, n$) as half of the relative mean difference. The mean difference is the expectation value of relative differences of randomly picked pairs from the set $\{x_1, \ldots, x_n\}$ and is computed as

$$MD(x_1, x_2, \ldots, x_n) = \frac{1}{n^2} \sum_{i=1}^{n} \sum_{j=1}^{n} |x_i - x_j| \tag{1.13}$$

Yes, this expression contains 0s for $i = j$ and all other values twice. The relative mean difference is the value normalized by the arithmetic mean of the same values. So, the Gini index is directly calculated as

$$\mathcal{G}(x_1, \ldots, x_n) = \frac{\sum_{i=1}^{n} \sum_{j=1}^{n} |x_i - x_j|}{2n \cdot \sum_{k=1}^{n} x_k} \tag{1.14}$$

For the case that all x_i are equal, obviously $\mathcal{G} = 0$. For the case that all wealth is in the hands of a single user (say user 1 who owns x), the Gini index computes to

$$\mathcal{G}(x_1, \ldots, x_n) = \frac{\sum_{i=1}^{n} \sum_{j=1}^{n} |x_i - x_j|}{2n \cdot \sum_{k=1}^{n} x_k} = \frac{2nx}{2nx} = 1. \tag{1.15}$$

So as before, if we want to use the Gini index to formulate our traffic rate assignment problem as optimization problem (here as a minimization problem) we face the problem that any assignment of equal rates to all users, no matter how small would yield a minimal Gini index of 0. And as before, it appears more reasonable to select the solution with the minimal Gini index among the efficient solutions only. This gives

$$\begin{aligned}
\mathcal{G}(x_1, x_2, x_3) &= \frac{2|x - (100 - x)| + 2|x - (100 + x)| + 2|(100 - x) - (100 + x)|}{2 \cdot 3(x + (100 - x) + (100 + x)} \\
&= \frac{2|2x - 100| + 200 + 4x}{6(200 + x)}
\end{aligned}$$

For $0 \leq x \leq 50$ this is equal to

$$\mathcal{G}(x_1, x_2, x_3) = \frac{200 - 4x + 200 + 4x}{6(200 + x)} = \frac{200}{3(200 + x)}$$

We can see that the Gini index becomes smaller for increasing x. Thus, the value is minimal for the largest possible $x = 50$. On the other hand, if $100 \geq x \geq 50$ the

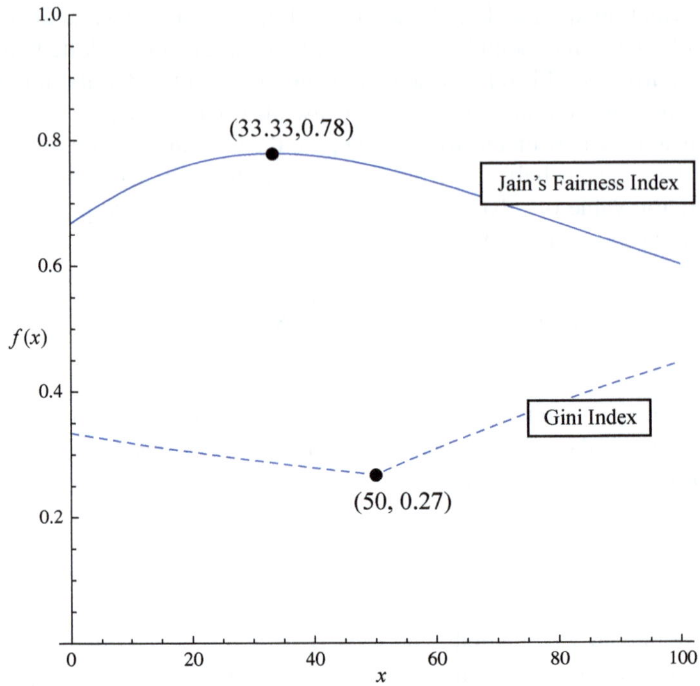

Fig. 1.2 Values of Jain's Fairness Index (solid line) and Gini index (dashed line) for Pareto efficient solutions of the assignment problem. The dots indicate extreme values of these indexes

Gini index becomes

$$\mathcal{G}(x_1, x_2, x_3) = \frac{4x - 200 + 200 + 4x}{6(200 + x)} = \frac{8x}{6(200 + x)} = \frac{4}{3(\frac{200}{x} + 1)}$$

Also, here the coefficient is monotone with x: for falling x, the denominator becomes larger and the Gini index smaller. Thus, it is smallest for the smallest possible $x = 50$. In summary, the Gini index is smallest for the assignment triple (50, 50, 150). We found entry number 5 in Table 1.1.

Figure 1.2 shows the change of both indices over the course of the Pareto efficient set. Despite the different locations of the extrema, there is no basic difference between the plots. The spread of both indices appears similar, as well as the rather smooth change, and the fact that there is only one extrema present. If a choice has to be made about which of the indices to prefer, there is not much to consider that distinguishes the indices. And it will not come as a surprise that there are a number of other indices that serve as measures for inequality and diversity (see, e.g., [24] for a number of other measures).

But there might be one observation that comes to mind, alas, a negative point—on first glance. The search for a Gini index of minimal value along the Pareto efficient set is somehow comparing apples to oranges. Taking the initial idea behind the Lorenz curve, it is the comparison of different ways of distributing the same total. Pun is on "the same." However, the Pareto efficient solutions all refer to varying totals. It can appear questionable to compare a way to distribute a total rate of 210 to another way of distributing a total of 270.

Moreover, once it is decided not to accept the maximum total solution since it includes a zero rate assignment, it means in clear-cut terms that the total of the solution that we decide at the end will definitely be less than the maximum (the total of 300 in our example case). So we can also change the perspective a little and assume that the total is set in advance, which can also be seen as an aspiration level. It can be the result of an admission control of the transport layer to pose such a condition on the total. Still, that target total is unknown, but while scanning over all feasible values for totals, we might get a sort of "optimality profile"—and the Lorenz curve will be of great help here.

But at first, bad news: if such a condition on the total rate exists. all feasible assignment will be efficient (Pareto efficient). It can be easily seen: for example, the condition is that $x_1 + x_2 + x_3 = 120$. Take any triple of rate values that fulfill this condition as well as the capacity constraints. Then any attempt to increase one of these values must result in the decrease of another value, simply because the sum of all values has to be preserved. So any such triple is efficient. This is also a demonstration that efficiency much depends on the feasible set. While the capacity constraints contain the feasible space such that for a given total (that must be at least 200) there is only one efficient assignment with such a total, if the feasible space is additionally constrained by a constant total rate, suddenly all assignments become efficient.

But in fact, from the admission control point of view only the total traffic within that network segment matters, but not how these rates are exactly distributed. Therefore, we can also understand optimality here to somehow bound the spread of the assigned rates. This kind of problem appears in many disciplines, and an important approach was devised nearly 100 years ago.

Assume that there are two ways of distributing the same total (wealth, commodity, whatever). If we plot both Lorenz curves, we can have a situation where the one curve does not cross the other. If the two sets of values are given as (non-negative) x_i and y_i (with i ranging from 1 to n), then $x_{(i)}$ should indicate the ith smallest value. We say "an" and not "the" ith smallest value since values can be equal. However, if there are multiple orderings of the elements in a non-increasing order, any of them can be selected—it will have no influence on the following exposition. And also, $y_{(i)}$ stands for the ith element of an ordering of the n y_i values in non-increasing order. Now the formal meaning of "not crossing" appears to be

$$\sum_{i=1}^{k} x_{(i)} \geq \sum_{i=1}^{k} y_{(i)} \text{ for all } 1 \leq k \leq n-1 \tag{1.16}$$

$$\sum_{i=1}^{n} x_{(i)} = \sum_{i=1}^{n} y_{(i)} \tag{1.17}$$

Table 1.2 Vectors of minimal majority for different values of the total C of assigned rates

Range of C	Minimal majority vector
$0 \leq C \leq 150$	$(C/3, C/3, C/3)$
$150 \leq C \leq 250$	$(50, 50, C - 100)$
$250 \leq C \leq 300$	$(C - 200, 300 - C, C - 100)$

It is said that x is majorized by y (referring to the fact that y has a larger spread). Behind this definition unfolds a beautiful theory with many cross-relations to other specifications—we will not go into detail here. But among these properties is one that says that if for two vectors one majorizes the other then also its Gini index will be larger or equal (this is the so-called Schur-convexity).

Now back to the example problem under the additional constraint that $x_1 + x_2 + x_3 = C$ where C is some fixed value between 0 and 300. Among all such assignments, we can look for assignments that are majorized by all other assignments, or that are not majorized by any other assignment. In a moment, we will see that both specifications make no difference here (they can make a difference in other circumstances). At the moment, we will also not care about how to find such assignments, but discuss the results shown in Table 1.2. In each specific case of C, we look for rates such that the vector composed of the smallest rate, the sum of the smallest and second-smallest (or median) rate, and C (as the sum of all three rates) is Pareto efficient—that is, there is no way to increase one of its components without decreasing any other. We call this vector ordered sum (by convention, we start with the smallest value despite the fact that ordering can be in two directions). It comes out that in each case we have an assignment that is majorized by all other feasible assignments, and which is not majorized by any other feasible assignment. Let us call this a vector of minimal majority (one can also say the "least spread" assignment). The structure of this vector depends on the ranges of values of C, in total, there are three cases.

Let us first confirm the results shown in this table by picking example cases that make the general pattern clear enough. In the first case, an equal rate assignment to all three users is possible. If the total assignment is more than 150, then it is not possible anymore to assign equal rates to all three users since that equal value would exceed 50 and thus violate the capacity constraint $x_1 + x_2 \leq 100$. The reason for the next border at 250 will be clarified in a moment.

Take, as an example, for the first case $C = 60$ (given that the total can be up to 300, this is a rather throttling demand). Then the minimum of the three rates (which is the first component of the ordered sum vector) is at most 20—otherwise the sum would exceed 60. On the other hand, the maximum is at least 20, and therefore $x_{(1)} + x_{(2)} = 60 - x_{(3)} \leq 40$. So the ordered sum vector for any feasible assignment such that the total is 60 is bounded from above by a vector $(20, 40, 60)$. But this is the ordered sum vector of a feasible vector: $(20, 20, 20)$. Thus $(20, 20, 20)$ is majorized by any other vector with a total of components of 60 and its components within the capacity constraints (1.1) and (1.2). So we have the equal assignment as the one with the least spread, where equal assignment is possible. And reminding Schur-convexity: it has the Gini index 0.

But when C exceeds 150, the equal assignment is no longer possible. In this case, the spread becomes influenced by the capacity constraints. From $x_1 + x_2 \leq 100$, actually two things follow: the first is that the minimum assignment can be at most 50. Otherwise, both the rate for user 1 and user 2 would be larger than 50 and their sum larger than 100. Then it is obviously $x_1 + x_2 \geq x_{(1)} + x_{(2)}$ since x_1 and x_2 are at least as large as the smallest and second-smallest value in the assignment. Therefore, the second component of the ordered sum vector is bounded by 100. The third is C. Thus, ordered sums are bounded from above by a vector $(50, 100, C)$, and again, this is by itself the ordered sum vector of the assignment $(50, 50, C - 100)$. This explains the second row in Table 1.2.

Now for the third row, where C exceeds 250. Why does the pattern $(50, 50, C - 100)$ not fit anymore? It is the second capacity constraint that snaps in: for $C > 250$, the value of $C - 100$ becomes larger than 150, and x_3 and x_2 together become larger than 200. Now take $C = 250 + m$ with $0 \leq m \leq 50$ and follow the reasoning.

$$x_{(1)} + x_{(2)} \leq 100 \tag{1.18}$$

$$x_{(3)} \geq 150 + m \tag{1.19}$$

$$x_{(3)} = x_3 \geq 150 + m \tag{1.20}$$

$$x_2 \leq 50 - m \tag{1.21}$$

$$x_{(1)} \leq 50 - m \tag{1.22}$$

Thus, the first component of the ordered sum vector is bounded from above by $50 - m$, the second (as before) by 100, and the third one is the pre-set total of $250 + m$. Switching back from m-notation to C-notation (where $m = C - 250$) it is the vector $(300 - C, 100, C)$. And again, this is the ordered sum of a feasible vector: $(C - 200, 300 - C, C - 100)$. We might not be immediately convinced that this is a feasible vector given the wanted ordered sum, so here is the check: the sum is $300 - C + C - 200 + C - 100 = C$. Then, for the first capacity constraint $x_1 + x_2 = C - 200 + 300 - C = 100$ and for the second $x_2 + x_3 = 300 - C + C - 100 = 200$—both are fulfilled, and we note in passing and with some good feeling that they even achieve the maximal values so that these assignments are also effective in the sense of not leaving resources unused. Last but not least: is the second component really the smallest? Yes, since $300 - C \leq 50$ while the other two components are at least 50. Then $x_1 < x_3$ and so the sum of smallest (x_2) and second smallest (x_1) becomes 100. The vector $(C - 200, 300 - C, C - 100)$ indeed produces the ordered sum vector $(C - 300, 100, C)$.

Figure 1.3 gives an impression of how differently the rates vary over the progress of an increasing total rate. Remember that the assignments were derived under the condition to spread assigned rates as low as possible (and by using majorization theory, this means that it holds for a broad range of possible interpretations of "small spread" by means of measures). The general picture of three different zones appears (indicated as I, II, and III in the figure), where the conflict between effectiveness and evenness is handled differently. In zone I—it might be called the "paradise"—equal rate assignment is possible, and this is exactly what has been done. With increasing

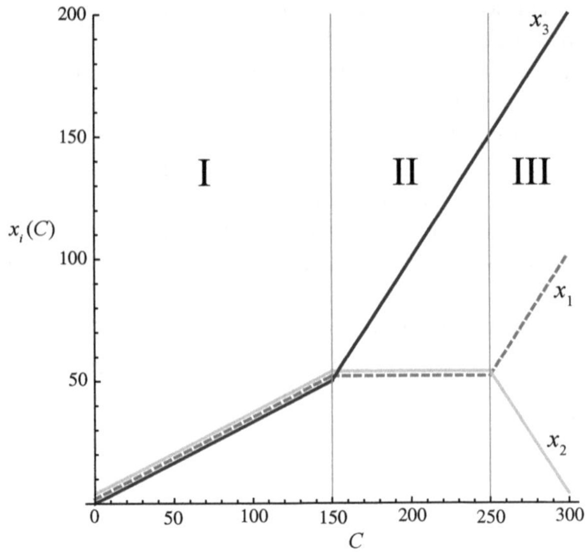

Fig. 1.3 Rate assignments of minimal majority for increasing values of fixed total rate (*Note* for better visualization, the curves are slightly offset against each other)

total, all rates grow linearly at the same rate. However, at one point, the rates for users 1 and 2 reach the first capacity constraint. It happens when the total is 150, and we reach zone II. Here it appears as if users 1 and 2 made a negotiation to keep their balance: each of them receiving 50 appears to be the best way to come along with the constraint that the sum of their rates must not exceed 100. However, user 3 does not feel attached to this contract, and further increase of the total rate goes directly into the wallet of user 3. Users 1 and 2 will not mind it since they are locked by the capacity constraint. All goes well as long as the total of user 2 and 3 does not violate the second capacity constraint of 200. This is at the border between zones II and III, and at this point, a number of relevant things happen: at first, the demise of user 2 starts, leading to the final rate assignment of 0 that we faced at the beginning of our exploration. Now we can better understand why this happens: any further increase in total traffic rate must be at the expense of balancing the impact of the capacity constraints. So user 1 quits the contract with user 2 to share the burden of their capacity constraint to equal parts, and user 3 quits the negotiation to grab higher rates only while not affecting the status quo between users 1 and 2. All this unfairness now is just caused by the larger supply of total traffic rates. Users 1 and 3 both can gain more by pushing user 2 to lower rates, even under the "best effort" promise to keep the imbalance of distribution as small as possible. It also means that at least user 1 can catch up with user 3.

So the first effect of overpowering the system is the inevitable demise of user 2's rate. The second is that from this point on, the least spread rate assignments become efficient. If we compare with our parametrized expression for Pareto efficient assignments ($(x, 100 - x, 100 + x)$ with $0 \leq x \leq 100$) we see that this fits with the expression for the rate assignments in zone 3 which is $(C - 200, 300 - C, C - 100)$, substituting $C = 200 + x$. The meaning of efficiency was that any attempt to increase the rate for one user must result in a decrease in the rate for at least one other user. Once the capacity constraints are reached (and all bottlenecks are filled), this can only mean that at least one user has to accept a decay. There is only one assignment of a total rate that is efficient and of lowest spread at the same time, and this is the assignment (50, 50, 150) (entry 5 in Table 1.1).

The next aspect is the difference between the global picture of assigning without reference to a specific total rate and the local aspect of assigning under a given total rate. The latter one serves a profile of optimal assignments that allows one to see the evolution of particular aspects of the global optimum much better. It might have come as a surprise that the optimal way of assigning such that the total rate becomes maximum leads to the exclusion of users. From Fig. 1.3 it appears to be the result of a process and the passing of a particular point where the total rate assignments cannot be achieved anymore in a fair manner.

Last but not least, following the process of increasing total rates, we see that within zones I and II the user rate assignments follow a rather simple growth pattern: rates grow equally for all users that contend for a remaining resource. Up to the capacity constraint of users 1 and 2, this is the equal distribution. After this, while entering zone II, users 1 and 2 do not (will not and cannot) contend for the remaining resource. But with a rate of 50, link 7 still provides a remaining capacity of 150, and only user 3 is left to take advantage of it. This goes up to the point when user 3 receives 150. After this, the growth of total rate cannot be continued anymore without violating the sharing principle that contended goods are shared equally.

We see that the optimality profile based on assignments of minimal majority gives a number of insights—but care has to be taken because so far we studied just a specific single example of a rate assignment problem. However, we will see in later parts that most of these aspects are of a general nature and apply to many more resource-sharing problems. At least two things should be kept in mind: a few moments before, we could not say which one of the efficient assignments is more favorable, the one maximizing the Jain's Fairness Index, or the one minimizing the Gini index. Now we have at least one more reason to favor the Gini index, since we can find the same assignment based on a rather different reasoning (but still with optimality in mind). The other lesson is that optimal solutions appear to be entangled in a conflict of efficiency and evenness. In this regard, zone II assignments are a trade-off between evenness with the probable waste of resources for the sake of maintaining indifference between users treatments, and unfairness resulting from a too large demand that cannot be well served by the distributing system due to its internal conditioning. It seems to be valid to call zone II the zone of fair assignments.

So far the taken point of view was that of an external deity that comes up with assignments based on some rationale that can be accepted by the users. So it can be the system provider, or a higher network layer framing. But the conflict that we wanted to handle with a suitable concept of optimality is with the user. So, how about letting the user decide for themselves how best to share the contested resource? This paves the way for an understanding of the system configuration as bargaining between users. What might not be expected at this moment, once taking this different path, is that actually we will again end up with the (50, 50, 150) assignment.

A game-like situation among bargainers was already studied by John Nash in the middle of the twentieth century [26]. In such a game, a number of bargainers contend about some commodities (note that in Nash's work two players were considered, but the analysis easily extends to any number of players. Note also that now we say "player" and not "user" but there is no pun intended...). There is an incentive to bargain. If the players fail to cooperate on a joint solution, all of them will receive nothing. If they cooperate, they can win. The theoretical analysis of Nash was focusing on four properties of a bargaining solution, and we will discuss these properties—then called "axioms"—here shortly. Reference is the so-called bargaining set, representing all possible ways to distribute a set of commodities based on player's utilities. So these sets have a numerical representation as a set of vectors from R^n (n is the number of players). We only consider convex and compact sets. The topological property of compactness stands for sets that are closed and bounded. Convexity stands for the property that if the set contains two points A and B then the set also contains all points on the line connecting A and B.

It is not so hard to see that our rate allocation problem can be seen as a bargaining problem with a compact and convex bargaining set. The feasible set of rate allocations is determined by the two bottleneck inequalities $x_1 + x_2 \leq 100$ and $x_2 + x_3 \leq 200$. Since all x_i are non-negative, this is a bounded set, and due to inclusion of the boundary cases, it is closed as well (it wouldn't be, for example, if the condition were $x_1 + x_2 < 100$ instead). Is it convex? If $A = (a_1, a_2, a_3)$ and $B = (b_1, b_2, b_3)$ fulfill these conditions, then for $0 \leq p \leq 1$ for the point on the line $\bar{AB}$ with coordinates $C = pA + (1-p)B$ will do as well, as can be easily seen. It is for the first condition

$$
\begin{aligned}
p(a_1 + a_2) &\leq p \cdot 100 \\
(1-p)(b_1 + b_2) &\leq (1-p) \cdot 100 \\
(p \cdot a_1 + (1-p) \cdot b_1) + (p \cdot a_2 + (1-p) \cdot b_1) &\leq p \cdot 100 + (1-p) \cdot 100 \\
c_1 + c_2 &\leq 100
\end{aligned}
$$

It can be shown for the second condition in the same way.

For the axioms, a general mapping is considered that selects from each bargaining set a single allocation (or distribution), the bargaining solution. Axiom 1 says that this selected solution should be Pareto efficient: if we have a solution on which all bargainers agree, then we can expect that there is no way that all bargainers can gain more at the same time. Axiom 2 is called impartiality, with the meaning that no external deity should influence the bargaining agreement and that the only information source is the bargaining set and nothing else (and not, for example, been influenced by the color of the hair of player 2). As a direct consequence, if all players have identical utilities and thus there is no source for differentiation among them, then the allocation should be identical for all players as well. Axiom 3 sounds innocent as well. It's called scale-independence. Assume that we obtain a new bargaining set (and related bargainer's problem) by scaling each player's utilities with some constant. The constant can be different for different players. The simplest example case here is that one player values in US dollars, another in euros. The one using the Euro now changes to the Chinese Yuan. After all, it should not influence the real value of the bargaining solution after the currency conversion. So, if we transform each element $u = (u_1, u_2, \ldots, u_n)$ of a bargaining set by multiplying component-wise with a vector $r = (r_1, r_2, \ldots, r_n)$ to yield $(r_1 x_1, r_2 x_2, \ldots, r_n x_n)$ at the end we will have a transformed bargaining set (we use u to indicate that the components are utilities for certain shares of the distributed commodities). Our bargain set solver will select a solution $s = (s_1, s_2, \ldots, s_n)$ from the original set and independently a solution $t = (t_1, t_2, \ldots, t_n)$ from the transformed set (the bargaining set solver does not need to know that the latter is a transformed version of the former). Then it must be for all i from 1 to n that $t_i = r_i s_i$.

We will stay for a while with that axiom before giving the last one. As the example of currency conversion shows, it is a natural requirement and fully competent to bear the name axiom. But coming back to our initial idea of allocating rates such that the sum of rates is maximized: using this as a bargaining solution, it would actually violate scale-independency—for the general case. A simple example, with the help of a software optimizer[1]: we consider non-negative x_1 and x_2 be bounded by a condition

$$x_1^2 + 2x_2^2 \leq 1 \tag{1.23}$$

(This is not the linear condition that we had so far, but it could be seen, for example, as an energy constraint). The solution with the maximum sum appears to be about $x_1 = 0.82$ and $x_2 = 0.41$. Now the scale of x_1 is changed by a factor of 2. The boundary condition changes to

$$\frac{1}{4}x_1^2 + 2x_2^2 \leq 1$$

and now the solution for maximal $x_1 + x_2$ is about $x_1 = 1.89$ and $x_2 = 0.24$. If it were scale-independent the solution should be $x_1 = 2 \cdot 0.82$ and $x_2 = 0.41$, which are different values. So we can keep in mind that scale-independency is not as self-evident as it seems on first glance, but also that the seemingly straightforward maximal-sum-approach has more than one drawback.

But there is one more axiom, which is actually the one that caused controversy later on. It is called "independence of irrelevant alternatives" and often abbreviated as IIA. It is a more far-reaching axiom than just for bargaining games. For example, one might think about the menu of a restaurant, offering the set of dishes that are served by that restaurant. Now, a guest makes a selection of her most favorable dish on a specific day. But at the same time, in a parallel universe, same restaurant, same customer, but the menu is shorter by a few dishes—luckily not the dish selected by our brave customer in our universe. IIA now means that the customer in the parallel universe would select exactly the same dish as the customer in our universe—the

[1] Mathematica 9.0.1 and its NMaximize function, to be precise.

missing choices of dish, as long as this does not affect the most favorable dish, will not affect the selection of the favorable dish in any way—they are "irrelevant alternatives" and the preference choice is supposed to be independent from their presence. Another example: we measure temperatures over a period of several days, and we find that last Friday was warmer than the Wednesday before. Now we continue our series of measurements, but no matter what we measure, the fact that last Friday was warmer than the Wednesday before will never change. It might seem obvious, but we will see that there are good reasons to have doubts about that principle. Even if it is owed to the fact that humans making decisions do not need to be rational all the time, our customer has a choice of cherry cake, strawberry cake, and cheese cake, and she chooses cherry cake. After placing the order, the waiter came back to the table and said that he is very sorry, but he had not given the choices correctly. In fact, cheesecake is out. "If that is the case, I will take strawberry cake instead of cherry cake."—this is a standard story illustrating what IIA truly means (see, e.g., [35]) but also how much we need to care about it.

Even then, Nash assumed that our bargainers are serious and that there should be an axiom representing IIA. But how would it appear in this case? It means that at first we do not have to solve a single bargaining problem, but provide a method to solve a number of bargaining problems in the same way. The meaning of "in the same way" is what Young calls consistency (instead of IIA) [40], and maybe it makes more sense to see it this way: if we happen to have the solutions for two problems that have a special aspect in common, there should be some common feature of the solutions as well. The link between two bargaining problems B_1 and B_2 should be: (1) the bargaining set of B_2 (i.e., the set of all possible outcomes of the bargaining) happens to be a subset of the bargaining set of B_1, and (2) even if it is a subset, it still contains the solution for the bargaining problem B_1—where both solutions were found by our common method. If both problems are linked this way, then the bargaining problem B_2 must have the same solution as the bargaining problem B_1. It means the removal of bargaining outcomes that haven't been selected as solutions anyway will not affect the solution of the smaller problem.

Here is not the right place to continue a discussion of this problem. What is important is that under these four axioms, Pareto efficiency, impartiality, scale-independence, and IIA, and the assumption of compact and convex bargaining sets, there is basically only one way to solve the bargaining problem: to maximize the product of the outcomes for all players.

The proof is actually not very complicated, and it has an interesting by-product that might help the player to understand why they should actually focus on the product maximization and not, for example, the maximum total (which we know would not necessarily be scale-independent). In the case of compact convex bargaining sets, the solution with the maximal product is also the solution where the total of relative changes for all players, compared to any other solution, will never be positive.

Let us see what this means for our allocation problem, assuming now that the three users who want to have a rate allocated to each and who are allowed to negotiate this between themselves. It means we want to maximize the product $x_1 x_2 x_3$ under our standard constraints. To find this maximum, we like to use a small trick by observing that an allocation with maximum product should be among the efficient allocations. Why? If it were not efficient, there would be another allocation with values of the same size or larger, and at least truly larger. The product is a strictly monotone operation, so this other solution would produce a larger product, and the former one would not be the maximum. Therefore, it must be an efficient solution, and we know already their general structure: choose $0 \leq x \leq 100$ and take the vector $(x, 100 - x, 100 + x)$. Now it only remains to find an x such that $x(100 - x)(100 + x)$ becomes maximal. After setting the first derivative to 0, solving the problem, and making sure that the result is larger than the boundary values, we find a solution

$$x = \frac{100}{\sqrt{3}} \sim 57.735 \tag{1.24}$$

So it is actually the "odd" solution that has been listed in row 6 of Table 1.1. It is not a comprehensive solution—but it has a remarkable property. If by whatever means our three bargaining users would have considered these solutions, they would get stuck at this point. For any other possible allocation, the total of relative changes would not be positive, or formally by denoting the product-maximizing solution as x^*.

$$\frac{x_1 - x_1^*}{x_1^*} + \frac{x_2 - x_2^*}{x_2^*} + \frac{x_3 - x_3^*}{x_3^*} \leq 0 \tag{1.25}$$

For example, instead of allocating $x^8 = (57.735, 42.265, 157.735)$, the users consider the (feasible) allocation $x = (60, 40, 150)$. Then user 1 gains about 4% while user 2 and 3 both lose about 5% and the total balance is -6%. The rule is that in no case will the total of all relative changes be larger than 0.

So if the users accept the relative change as a standard of comparison for all other possible allocations (which appears to be more tangible than the goal of jointly maximizing a product since it directly refers to perceivable wins and losses) they will not be able to find any other solution where at least all can win in total. So there is no incentive to consider any other solution.

However, as perfect as it seems, there is an obvious obstacle: the solution is simply odd. Assume our bargainers meet in a hut deep in the mountains, without any access to the internet or any computing device, would they ever consider this odd solution? Critique of the Nash solution came actually from the experimental side, owing to the fact that such a kind of matchmaking is hardly observed in reality. One alternative solution, better reflecting human decision-makers, was provided by Kalai and Smorodinsky in [19] by swapping the IIA requirement with a so-called monotonicity axiom. Without going into to many details now, this axiom reflects another kind of consistency between bargaining solutions for bargaining problems that are linked. It considers the functional dependency of the maximum that one player can achieve when the outcomes for all other players are kept constant. If for one bargaining problem this function appears generally larger than for another problem, then also the outcome of the solution for the corresponding player should be larger. Some consideration gives (see, e.g., [40]) that also in this case, there is basically only one solution, and it can be constructed in the following way.

1. For each player (bargainer, user) $1 \leq i \leq n$ find the maximum outcome p_i that she could get over any possible outcome.
2. Find the largest r such that the vector $(rp_1, rp_2, \ldots, rp_n)$ is feasible.

In fact, we can solve this procedure easily for our studied allocation problem. From $x_1 + x_2 \leq 100$, both users 1 and 2 could get at most 100. For $x_2 + x_3 \leq 200$ user 3 could get at most 200. Then, what is the largest r that $(100r, 100r, 200r)$ is still within these constraints? Again, we experience the "bottleneck" to come from the first condition: in such an allocation users 1 and 2 will get the same rate, and thus r cannot be larger than 0.5. Then, the allocation for $r = 0.5$, which is (50, 50, 100) is feasible since it fulfills both constraints and we have found the Kalai-Smorodinsky bargaining solution. It might, in fact, be acceptable to all parties that all get the same share of their maximum possible shares, and that this result cannot be accomplished on a higher level. But as we can easily see, this solution is not efficient anymore: one user can get more without reducing the rates of any other user. In this case, user 3 could receive higher rates than 100 without affecting the allocation to users 1 and 2 in any way.

However, the product-maximizing solution gave raise to another generalization. We will now leave the bargainer's model and come back to the plain problem of assigning rates to users, having their consent or not. One observation is that the product-maximizing solution (later on we will call it the proportional fair solution) appears rather close to the (50, 50, 150)-solution that we have faced now already several times. Is this just a random coincidence? To see that it is in fact not a coincidence, we will start with a special "interpretation" of "maximizing a product"—based on the fact that this corresponds with maximizing the sum of the (natural or any other base) logarithms of the allocated rates. This is because the logarithm is a strict monotone function and we have

$$\max\left[\prod_{i=1}^{n} x_i\right] \equiv \max\left[\sum_{i=1}^{n} \log x_i\right]. \tag{1.26}$$

Now, for some reason, the logarithm is a welcome mediator between a value and a prospective utility of a user for such a value. It means the user values rates assigned to her not simply by their numerical value but by the degree to which the background technical infrastructure allows her to fulfill her needs. The economy has considered a so-called "law of diminishing returns" for a long time, especially with regard to the case that she would receive a huge rate—a rate so huge that a reasonable utilization would not become possible anymore. Logarithm is commonly seen as a convenient way to alleviate this effect. More on this later. For now, we keep in mind that the max-product approach, as well as its "sister approach" to seek a state where not all users can gain a total win in relative changes—can also be seen as a utility-maximizing approach, at least as a special case of it.

In that regard, if the utilities were to just equal the values, then we would end up with a solution that we had already at the beginning, and where we haven't been so happy about the outcomes: to maximize the sum of rates. But how about other ways of valuating an assigned rate?

The term rate already reflects the potential transmission of an amount of data per time unit (like "100Mbit per second"). If the rate is being symbolized as x, then $1/x$ refers to the time it takes to finish a transmission of a unit amount of data, or potential delay in other words. Note that there is no simple relation between an assigned rate and the time it takes to send a complete file, as this process is much influenced by the various protocol issues within the various network layers. But no doubt that on the first hand it will be an inverse-proportional relation. So the utility of a rate x would become $1/x$ and we use it instead of $\log x$ to meet another way of expressing our optimality concern of the rate allocation. But now we have to seek a minimum, as delay (waiting time) is something that feels better when it is small.

$$\min\left[\sum_{i=1}^{n}\frac{1}{x_i}\right] \tag{1.27}$$

Naively, one would not expect a totally different solution from the task of maximizing the sum. But it is different. Also, here we can use the "Pareto trick" since increasing all rates would raise the value of the sum of inverse rates, and also here, the minimizing rate allocation must be among the efficient solutions. So we would have to find the extrema of the function

$$f(x) = \frac{1}{x} + \frac{1}{100-x} + \frac{1}{100+x}, \quad \text{for } 0 \le x \le 100. \tag{1.28}$$

The analytical form of the solution is rather complex; we take a rounded numerical value

$$x = 51.362 \tag{1.29}$$

Therefore, the optimal rate allocation for minimizing potential delays would be $(51.362, 48.638, 151.362)$. There are two immediate observations:

1. This solution for minimizing the total of inverse rates is very different from the solution for maximizing the total of rates, especially since the former does not show the tremendous unfairness of the latter case.
2. It comes even closer to the (50, 50, 150) solution than the maximum-product solution.

The first observation makes some trouble for a purely utility-based point of view on group decision making. In fact, why should the "comfort" of a larger commodity *rate* be something totally different from the "discomfort" of a larger commodity *potential delay* (or: waiting time)? It seems hard to explain... But by second observation we already gave a bias to the interpretation that in fact a series exists that converges to the (50, 50, 150) solution (yet without giving a name to this solution). Instead of a spurious utility that becomes more and more vague on each further argument, there is in fact a more simple explanation for the appearance of a series. And we can add that also the maximum total case follows this pattern, as we observe something like

$$\text{maximum sum} > \text{maximum product} > \text{minimum sum of inverted} > (50, 50, 150). \tag{1.30}$$

We see that this series includes the three most common *power means*, the arithmetic, geometric, and harmonic means. The general qth power means for a series $x = (x_i)$ $(i = 1, \ldots, n)$ of n non-negative data is defined as follows.

$$P_q(x) = \left[\frac{1}{n} \sum_{i=1}^{n} x_i^q \right]^{1/q} \quad \text{where } q \in I \setminus \{0\} \tag{1.31}$$

for integer $q \neq 0$ and

$$P_0(x) = \left[\prod_{i=1}^{n} x_i \right]^{1/n} \tag{1.32}$$

for $q = 0$ (the geometric mean). What matters more here is a well-known general inequality among power means. For $p < q$ we have

$$\min[x] \leq P_p(x) \leq P_q(x) \leq \max[x] \tag{1.33}$$

where equality only holds if all x_i are equal (and remember non-negative).

We see a similar relation for the maxima of three power means in the analysis—but generally there is no reason to assume that the power means inequality transits to the power means maxima as well, and also, we did not clearly specify what we understand by "distance" to some feasible point like the $(50, 50, 150)$-allocation. But what is relevant is that everything is lower-bounded by the minimum.

At the beginning of this chapter, we started to compare rate allocations by the total rate assigned to all users. This is obviously the same as maximizing the average rate that each user receives. When studying bargaining allocations, we came to the point of maximizing the product among all feasible allocations. Also here, we can see it as a case of maximizing the geometric means (we have just to take the nth root of the product, which is a monotonic operation). Last but not least, minimizing the total potential delay appears to be the maximization of the inverse of the average potential delay per user—and this is exactly the maximization of the harmonic means. So we are back to a maximization formulation as well. Now its natural to consider maximization of further power means where $q < -1$. It might lack a rationale like "potential delay" for $q = -1$ or so, but nevertheless, it's a reasonable question to ask where we might end up with such a process.

As a matter of fact, we will come close to some concept of "maximizing the minimum" but with a more specific meaning. To put the question in a more formal sense: for each possible q we maximize $P_q(x_1, x_2, x_3)$ and then let q go to $-\infty$—is this process converging?

It will indeed converge, and even more, it will converge to that (50, 50, 150)-allocation. It is not an obvious fact, and for now we state only that a series of comparisons between two vectors x and y by qth power means will approach a so-called lexicographic maxmin comparison. It means that we sort the components of both x and y in non-increasing order (if these vectors contain equal components and thus there are multiple ways of sorting, then there will be no influence on the following procedure). Then we compare both vectors in a lexicographic sense, i.e., like sorting the books in a library starting with books where the titles start with letter "A," then "B," etc. If two book titles start with "A" we compare by the second character, etc. We sort like this with the main difference that we start with the largest character, not the smallest. So first come all vectors with the largest smallest component. Among all of them, first come the vectors with the largest second-smallest component, etc. If we think of all feasible states in our allocation problem, sorted that way, which one will come first? Yes, its the (50, 50, 150)-allocation.

It is not so hard to see in this special case. The inequality $x_1 + x_2 \leq 100$ can be also written as $\min[x_1, x_2] + \max[x_1, x_2] \leq 100$ and since $\min[x_1, x_2] \leq \max x_1, x_2$ we have $2\min[x_1, x_2] \leq 100$ or $\min[x_1, x_2] \leq 50$. The choice $\min[x_1, x_2] = 50$ is feasible and we have allocations where $x_1 = 50$ causing also $x_2 = 50$ and vice versa. So all that remains is to decide about the value of x_3. It is a decision among all allocations $(50, 50, x_3)$ where $x_2 + x_3 = 50 + x_3 \leq 200$ or $x_3 \leq 150$, and the largest possible third-smallest value in our allocation will be 150. So the allocation (50, 50, 150) has the largest "lexicographic value" and is at the same time an allocation that maximizes the minimally assigned rate—but with regard to the latter specification, by far not the only possible allocation doing so.

But now back to the maximization of power means, we would like to ask as well: how about related "standards of comparisons"? Are there such SoCs in the case of power means, and do they converge as well? This goal can be achieved when we consider a construct of the formal form "rate of rate of rate of ... rate of relative change." We had before the relative change as SoC for maximizing the product, which means (1) that a transfer is accepted when the total of relative changes is positive, and (2) that we found an allocation of rates where there is no acceptable transfer to any other allocation possible. Then studying potential delays in fact it came out as maximality of the harmonic means—and the SoC here is the rate of relative change, or—so to say—the relative change. So it refers to the fact that a relative change of 10% can mean something different for a millionaire or a poor person. After an exemplary 10% relative loss, the millionaire might still be able to have caviar and egg for breakfast, while the poor person might have trouble having breakfast at all. So we weight the relative change by the total wealth once more, and consider the net total in the way

$$\sum_{i=1}^{n} \frac{1}{x_i} \frac{y_i - x_i}{x_i} \leq 0 \tag{1.34}$$

which can, of course, also be written as

$$\sum_{i=1}^{n} \frac{y_i - x_i}{x_i^2} \leq 0 \tag{1.35}$$

We come up with a concept generally known as α-fairness. More on this later, for now, we consider the specific definition of a way of comparing two allocations.

$$\sum_{i=1}^{n} \frac{y_i - x_i}{x_i^\alpha} \leq 0 \tag{1.36}$$

where we have seen that $\alpha = 1$ corresponds with the maximum product (or maximum total log utility) and $\alpha = 2$ with the potential delay, or maximum harmonic mean, or maximum total $1/x$-utility (but why should someone utter such a utility?). Then for each $\alpha \geq 1$ we can find an allocation of rates such that any other allocation fulfills the condition like in the inequality of Eq. 1.36—meaning that a transfer to any other allocation is not acceptable and therefore some good reason to prefer this allocation—for each α.

Do we have a SoC for the lexicographic maxmin sorting as well? Actually, this is a little bit more intricate a question, but there is a rough answer. At least in the case of our allocation example, it works fine and goes like this: for any other allocation, if we try to increase the rate for one user, there will be another user who must receive less (otherwise we leave the range of feasible allocations). So far, it is just repeating Pareto efficiency. But that user who receives less happens to be a user who does not already have more than the user with the increased rate. So the "Robin Hood" sort of theme here is to take from the poor to make the rich more rich. And we are looking for an allocation to which a transfer to any other allocation where someone receives more would always need to make a poor more poor.

We can see that the (50, 50, 150) allocation has this property. We try to increase the rate for user 1 beyond 50. Since $x_1 + x_2 \leq 100$, it means we must reduce the rate of user 2 below 50. So user 2 is such a comrade that had, before the change, the same as user 1 and now less. The same happens if we try to increase the rate for user 2 where we cannot avoid to take from the same-rate user 1. Finally, if we try to increase the rate for user 3 we must still ensure $x_2 + x_3 \leq 200$, and it would need to reduce the rate for user 2. But for user 2 the allocation 50 was already lower than the rate of user 3 of 150. The reader is also invited to note that among all feasible allocations, there is no other allocation with such a property, i.e., if we try to increase one rate, we always have to reduce another rate for a user that has already the same or less.

Now we had a few times a ("Rawlesian") hint on considering the situation of the user with the lowest rate assignment. Then how about closing the circle and coming back to the initial idea to maximize the total of rates—but now we give a little bit higher weight to users with smaller rates. Since we cannot know the formal indices of users with low rates, we use a new notation: by $x_{(1)}$ we indicate a user with the smallest rate ("a user" and not "the user" since there can be more than one...). In a similar way, $x_{(2)}$ is a user with the second-smallest rate, and last but not least, $x_{(3)}$ is a user with the maximum rate—we have only 3 users in our example, so it coincides with any user with the third-smallest rate. Using this notation, we ought to maximize

$$4x_{(1)} + 2x_{(2)} + x_{(3)} \tag{1.37}$$

instead of $x_1 + x_2 + x_3$ (that we can also write as $x_{(1)} + x_{(2)} + x_{(3)}$). What do we get? Let us first look at a base point for the value of the (50, 50, 150) allocation. It is $4 \times 50 + 2 \times 50 + 150 = 450$. Can we find an allocation with a larger value? To solve such optimality problems—for now—we have to study the various possible orderings of the allocation. At first, we can see that x_3 cannot be the user with the smallest allocation. Since at least one of x_1 or x_2 must be smaller than or equal to 50, it implies two things: $x_3 \leq 50$ as well (it is our minimum and must be smaller than the minimum of x_1 and x_2) and $2x_{(2)} + x_{(3)} \leq 150$ since this is taken by $x_1 + x_2 \leq 100$ plus the smaller one of x_1 and x_2 that is at most 50. So putting both together, whenever x_3 is the minimum rate, the total of our weighted expression cannot exceed $4 \times 50 + 150 = 350 < 450$.

Now consider that x_1 is the minimum. Then either x_2 or x_3 is the second-smallest value. In the former case, i.e., where we have the order $x_1 \leq x_2 \leq x_3$ and assuming the minimum $x_1 < 50$, we can write the weighted total as

$$\underbrace{x_1 + x_1 + x_1}_{<150} + \underbrace{x_1 + x_2}_{\leq 100} + \underbrace{x_2 + x_3}_{\leq 200} \tag{1.38}$$

and the whole sum would be less than 450. But if $x_1 = 50$ it implies $x_2 = 50$ and $x_3 \leq 150$. The expression will be maximal for the final choice $x_3 = 150$, and we are back to our (50, 50, 150) allocation.

In the latter case (x_1 is minimal and x_3 is the second smallest), it goes similarly. We only have to take note that x_3 being the second smallest implies that it must not be larger than either x_1 or x_2 and thus cannot exceed 100.

$$4x_1 + 2x_3 + x_2 = \underbrace{x_1 + x_1 + x_1}_{<150} + \underbrace{x_1 + x_2}_{\leq 100} + \underbrace{x_3 + x_3}_{\leq 200} \tag{1.39}$$

as long as $x_1 < 50$, and also here raising it up to the maximal value 50 guides to the (50, 50, 150) solution.

The other choice is x_2 as the minimum. Now assuming that x_1 is second smallest and $x_2 < 50$ we can see

$$4x_2 + 2x_1 + x_3 = \underbrace{x_2}_{<50} + \underbrace{2x_2 + 2x_1}_{\leq 200} + \underbrace{x_2 + x_3}_{\leq 200} \tag{1.40}$$

and the total cannot exceed 450. Raising x_2 to its largest possible value (before x_1 would take over to be the minimum value) again gives the maximum value for $(50, 50, 150)$. In the final case, x_3 is the second smallest, and also here it would be at most 100 since it has to be smaller than or equal to x_1. We continue

$$4x_2 + 2x_3 + x_1 = \underbrace{x_2 + x_2 + x_2}_{<150} + \underbrace{x_2 + x_1}_{\leq 100} + \underbrace{x_3 + x_3}_{\leq 200} \tag{1.41}$$

and the sum cannot exceed 450. As before, it will be maximal for $x_2 = 50$ and guiding again to the $(50, 50, 150)$ allocation.

With some effort, we could solve to maximize this weighted total, and we come up again with the $(50, 50, 150)$ allocation. In other words, this allocation maximizes 4 times the smallest plus 2 times the second-smallest plus the largest allocation value.

Actually, the weight 4 for the user with the minimum rate was a kind of educated guess. If we had taken only 3 as the weight value, it would look different. Here the $(50, 50, 150)$ allocation gives $3 \times 50 + 2 \times 50 + 150 = 400$ but the allocation $(100, 0, 200)$ maximizing the total $x_1 + x_2 + x_3$ would have the same value: $3 \times 0 + 2 \times 100 + 150 = 400$.

Starting with a simple and innocent question about the best way to allocate rates to users in a so-called end-to-end traffic scenario, we could come up with a number of solutions. After all, each one is "best" by itself, and without further qualification, there is no way to prefer one over the other. However, there might be two different conclusions at the end:

Conclusion number 1 is the opportunist: haven't we seen the rate allocation 50 to user 1, 50 to user 2, and 150 to user 3 now so many times appearing from completely different contexts? Then it's time to declare it the winner and a way of allocating rates that serves best in many regards.

Conclusion number 2 is conservative: it's better to avoid sharing of links while establishing a routing, for example, by drawing new connections.

Chapter 2
Computational Fairness in Resource-Sharing

In the foregoing chapter, we have extensively studied a single and rather simple example of an allocation problem. We have seen a number of ways to specify "optimality" in order to solicit a single solution among all feasible allocations. It was also observed that some relations appeared between these different ways and their solutions, leading to multiple appearances of the same solutions. But we haven't seen much about generality. It could still be just a random coincidence. We have to consider something usually called the general case. But what generalization applies here?

In the next few lines of text, we will show that all optimality concepts we have seen so far can be put into a simple and general framework based on four definitions. We will give these definitions now.

Definition 2.1 (*Binary Relation*) A *binary relation* R over a domain A is a subset of $A \times A$.

Definition 2.2 (*Greatest Element*) A greatest element of a relation R over domain A is any element $x \in A$ such that for all $y \in A$ $(x, y) \in R$.

The set of all greatest elements is called the *best set* of the relation R over the domain A.

Definition 2.3 (*Asymmetric Part*) The *asymmetric part* $P(R)$ of a relation R over domain A is the set of all pairs $(x, y) \in R$ where $(y, x) \notin R$.

We note in passing that the remainder of the relation set of pairs, i.e., $R \setminus P(R) = I(R)$, is called *symmetric part*. If someone wonders about the use of letters P and I: this notation is taken from social choice theory and P stands for *Preference* while I stands for *Indifference*.

Definition 2.4 (*Maximal Element*) A *maximal element* of a relation R over domain A is any element $x \in A$ such that for no $y \in A$ $(y, x) \in P(R)$.

M. Köppen et al., *Resource-aware Intelligent Communication Systems*, Springer Tracts in Human-Centered Computing, https://doi.org/10.1007/978-3-032-06915-3_2

The set of all maximal elements is called the *maximum set.*

If we call the greatest and maximal elements together as extreme elements, then the bottom line of all approaches that were discussed in the foregoing chapter is that optimality always appears as an extreme element of some relation over some domain. We did not study different "optimalities"—we studied different relations and their domains.

Let us reconsider two cases from the foregoing section to see the definitions in action. Just to recall, we were considering a system where traffic rates x_1, x_2, x_3 had to be allocated to 3 users under the link capacity constraints $x_1 + x_2 \leq 100$ and $x_2 + x_3 \leq 200$. We had the case of maximizing the total rate allocation $x_1 + x_2 + x_3$. In fact, the link capacity constraints define a subset of R^3_+ that can serve as the domain A of a relation. Now, taking any pair (x, y) of vectors from this domain A, we can logically decide whether they belong to a relation R by comparing the sum of components. In case the sum of components of vector y is not larger than the sum of components of x, and only then, we say that the pair of vectors (x, y) belongs to our now defined relation R. So, for example, take $x = (100, 0, 200)$ and $y = (50, 50, 150)$. The sum of components of x is 300, and the sum of components of y is 250. As $300 > 250$ it confirms that the pair $((100, 0, 200), (50, 50, 150))$ belongs to our relation. Simply, a pair (x, y) belongs to R if and only if

$$(x, y) \in R \leftrightarrow x_1 + x_2 + x_3 \geq y_1 + y_2 + y_3 \tag{2.1}$$

Now, when asking for an allocation that maximizes $x_1 + x_2 + x_3$, it's the same as asking for a rate allocation such that every other allocation has a sum of components not larger than the former's sum of components. If we denote the seeked "optimum" allocation (i.e., the one with maximal sum of components) as x' and any other feasible allocation by x (i.e., any $x \in A$), then it means exactly what our relation defines: the pair (x', x) must belong to R. So formally, for all $x \in A$ we have $(x', x) \in R$. This is exactly what Definition 2.2 specifies, i.e., x' is a greatest element of the relation R over domain A.

This is a quite common and convenient way to specify optimality: for any pair, compute some numerical value for each element of the pair and see if the value for the first is larger than or equal to the second. Depending on the circumstances, this "scalar" function (as we map domain elements into R^1) is called quality function, coast function (here it should be as small as possible), fitness function, or just aggregate function. In any case, its a function. We have seen several other cases of such scalar functions in the foregoing chapter, for example, the Gini index, Jain's Fairness Index, the product, any power means, or the "weighting after sorting" approach at the end.

On the other hand, in the foregoing chapter, the so-called Pareto efficiency was considered. Here, we have compared two vectors of rate allocations in the most canonical way. The vector x is seen in relation to another vector y if for all component indices i we have $x_i \geq y_i$. Thus, a rate allocation $x = (50, 50, 150)$ would be in relation to the rate allocation $y = (50, 50, 100)$ since each component of x is at least as large as the corresponding component of y.

What is the asymmetric part of this relation? Again, we prefer to identify the symmetric part and thus we have to look for all pairs (x, y) that belong to the relation together with their converse (y, x). Formally, it means that for each component index i we must have $x_i \geq y_i$ as well as $y_i \geq x_i$. This is only possible if for all i $x_i = y_i$ and it means $x = y$ (such relations are called antisymmetric relations, but more on this later). So the asymmetric part is the full relation itself, except all pairs (x, x) where $x \in A$. This relation is known under the name Pareto dominance Relation, sometimes also simply "vector inequality" and is formally defined as follows.

Definition 2.5 (*Pareto Dominance Relation*) Given a domain $A \subseteq R^n$ and two elements $x, y \in A$. It is said that x Pareto dominates y (also written as $x >_p y$) if and only if for all component indices i it is $x_i \geq y_i$ and for at least one component index j it is $x_j > y_j$.

The maximal elements of A are now vectors where it is not possible to increase all their components at the same time. It means for any such maximal x' there is no other vector $x \in A$ with $x >_p x'$. At least one component must decrease, and this implies that the maximum set is actually the set of all Pareto efficient solutions: for each increase, there must be at least one decrease.

Let us define a few other means of comparison that we have met in the foregoing chapter as corresponding relations. Now we also introduce the names under which they are known and commonly studied.

Definition 2.6 (*Proportional Fairness*) Given a domain $A \subseteq R^n_+$ and two elements $x, y \in A$. The pair (x, y) of elements from A is said to be in *proportional fairness relation* if and only if the following inequality holds.

$$\sum_{i=1}^{n} \frac{y_i - x_i}{x_i} \leq 0 \tag{2.2}$$

Definition 2.7 (*Maxmin Fairness Relation*) Given a domain $A \subseteq R^n$ and two elements $x, y \in A$. The pair (x, y) of elements from A is said to be in *maxmin fairness relation* if and only if for all component indices i with $x_i < y_i$ there is at least one component index j such that (1) $x_j \leq x_i$ and (2) $x_j > y_j$.

For the following definition, we recall the notation $x_{(i)}$ for an ith smallest component of the vector x. Note that there might be equal components and thus different ways of sorting the components, so if using such notation, it has to be ensured that this ambiguity does not influence the results.

Definition 2.8 (*Leximin Relation*) Given a domain $A \subseteq R^n$ and two elements $x, y \in A$. The pair (x, y) of elements from A is said to be in *leximin relation* if there is a component index k such that for each $i < k$ it is $x_{(i)} = y_{(i)}$ and $x_{(k)} > y_{(k)}$.

This relation also appears under other names: the most correct name is lexicographic maxmin relation, or LMM, but also lexmax is possible.

Another notation: the *ordered sum* $O(x)$ of a vector x is the vector composed of the smallest, the sum of the smallest and second smallest, the sum of the three smallest up to the sum of all components. By $O(x)_i$ we denote the ith component of such a vector. As an example: if $x = (50, 50, 150)$ then $O(x) = (50, 100, 250)$.

Definition 2.9 (*Majorants Relation*) Given a domain $A \subseteq R^n$ where for all $x \in A$ $O(x)_n$ has the same value. The pair (x, y) of elements from A is said to be in *majorants relation* if and only if for all component indices i $O(x)_i \geq O(y)_i$.

2.1 A Primer on Binary Relations

We have already seen the basic definition of a binary relation. It is only using basic concepts of set theory: the meaning of a set itself, the direct product of sets, and the concept of a subset. Thus, binary relations are a very general concept that can be applied in a multitude of contexts.

The domain of a relation can be any set—with the exclusion of "sets of all sets" as it is common in set theory. We have only seen relations with a continuous domain so far, or even more specifically with a domain that is a subset of some n-dimensional vector space. However, the domain can also be a discrete set, and it can be finite.

According to Definition 2.1, a binary relation is given as a set of pairs. In practical circumstances, it might be inconvenient to handle such a list, virtually impossible if the domain is infinite or continuous. Luckily, we can represent relations in various other ways.

1. **Point-wise representation**: it is just another name for set representation.
2. **Predicate representation**: (also *logical representation*) we give the logical condition whose truthfulness decides whether a pair of domain elements belongs to the relation or not. This might be the most common way of specifying a relation in practice since it is most able to express the concept, rationale or idea behind a relation. For example, we take the relation $\{(5, 3), (3, 1), (5, 1)\}$ over the domain $\{1, 3, 5\}$. The condition for a pair (x, y) of domain elements to belong to that relation is simply $x > y$ so the relation is in fact a value comparison. However, the predicate representation can easily make forgetting the domain issue. Sometimes care has to be taken that any conclusion drawn from the predicate for one relation might not be true any longer if taking the same predicate but a different domain. Especially, the transition from continuous to finite domains can be easily affected by this problem.
3. **Functional Representation**: We map each domain element to the set of domain elements to which it is in relation. So we have the mapping $f : A \rightarrow 2^A$ where from $y \in f(x)$ follows $(x, y) \in R$. Especially in the continuous case, this representation is useful, for example, for visualization purposes when we can get an overview of which domain elements are "dominated" by some given x according to a given relation. Taking up the example for the predicate representation, the functional representation of the same relation would be the mapping $f(1) = \emptyset$, $f(3) = \{1\}$ and $f(5) = \{1, 3\}$.

4. **Graph representation**: In the case of a finite domain, the relation can also be represented as a directed graph. Each domain element represents a node, and two nodes x and y are connected by an arc (or directed edge) from node x to node y if and only if (x, y) belongs to the relation. Graphs are useful for understanding global properties of relations. For example, the concept of a cycle of a relation refers to a cyclic path in a directed graph. In this matter, we might like to define a directed graph with a continuous set of nodes by means of a binary relation in any of the representations above.
5. **Matrix representation**: also here it needs a relation over a finite domain. Since in this case it can be represented by a directed graph, it is well known that the directed graph, in turn, can be represented as a matrix with entries 0 or 1. But we need to impose an index order on the domain elements. Thus, we have to use a notation like x_i to indicate the ith element of the domain. For such a matrix M $M_{ij} = 1$ if and only if the pair (x_i, x_j) belongs to the relation. Otherwise, the entry is 0. Generally, this matrix representation is not of big interest and one will not meet it often.

Since relations are sets by themselves, they can be treated like sets, with regard to set union, intersection, negation, etc. But we also want to mention domain (domains-as-a-set) manipulations. If we take any relation and restrict its domain somehow, we get a new relation that only contains pairs where both elements still belong to the new domain. It means when two people meet on the street and start to chat about "sub-relations" they have to make sure that they speak about the same thing. The word "sub-relation" can refer to a subset of the set of pairs, or a subset of the domain. But in general, as a relation-specific term, it refers to the subset of the domain, so we have a definition:

Definition 2.10 (*Sub-Relation*) Given a relation R over the domain A. A *sub-relation* of $R|_B$ over a domain $B \subseteq A$ is the set of all pairs $(x, y) \in R$ where $x, y \in B$.

It means that there are pairs where at least one element does not belong to the new domain set is dropped. If we take our small example from above: if the domain is restricted to the set $\{1, 5\}$, then the new relation is given by the set of pairs $\{(5, 1)\}$. Of course, a sub-relation is also a subset of the relation (as a set of pairs)—just there is a certain specification of that subset.

We will not go into much detail, but it has to be mentioned that relations are a rather "deep" concept of mathematics, and they intermingle with a number of other concepts. If we take the functional representation of a relation and each mapped element contains at most one element, then the relation is called a function. On the other hand, the functional representation specifies a relation as a special function. So it's a mutual specification of functions from relations and relations from functions. The other concern has to be given about a special relation that is usually banned from further consideration, as it would refer to a "set of all sets": the subsethood relation. It can only be defined for a specific power set of some set (since this is a "normal" set), otherwise we could face mathematical conundrums that are better avoided.

For each (valid) domain A there are three special relations that can always be defined.

1. The empty relation $\emptyset_A$ (or just $\emptyset$), which is just a relation that contains no pairs. It is reasonable to have such a relation, for example, if its logical representation is based on a criterion that is never fulfilled in the specific circumstances.
2. The full relation U_A that is equal to the direct product of A with itself, or: $U_A = A \times A$. It means a relation where any possible pair is in relation to each other. It's something like a fully connected directed graph. Another name is *universal relation*.
3. The *diagonal relation* D_A that contains all pairs (x, x) where $x \in A$, and only these. It's called diagonal by reference to the matrix representation (which, for a finite domain, would be a diagonal matrix in this case). Another possible name is *identity relation*.

Next, we will look for specific properties of relations. In the following, R indicates the relation, A its domain, and x, y, z are domain elements (not necessarily different ones).

1. A relation is called *reflexive* if for any $x \in A$ $(x, x) \in R$. We can now also say that a relation is reflexive if it contains the diagonal relation over its domain. Reflexivity is not a strong demand, as any relation can be easily patched to make it fully reflexive or not reflexive at all.
2. If we have for any two domain elements x and y that $(x, y) \in R$ or $(y, x) \in R$ (or both) then we call the relation *complete*. Since we can also pick x two times, such relations must be reflexive as well.
3. A relation is called *symmetric* if from $(x, y) \in R$ follows $(y, x) \in R$. Remember that we have already mentioned the symmetric part of a relation. So the symmetric part is a symmetric relation. This property refers to a relation where the order in the pair (x, y) does not matter. Thus, the domain elements x and y appear to be equal, similar, close, etc.
4. We saw already the definition of the asymmetric part. A relation is called asymmetric if from $(x, y) \in R$ strictly follows $(y, x) \notin R$. Yes, the asymmetric part of a relation is an asymmetric relation. Asymmetry is a clear indication that the order of elements in the pair (x, y) matters and that we are about to make a comparison between elements, or to order them under some aspect.
5. It sounds very similar, but is a different property of a relation: a relation is called antisymmetric if from $(x, y) \in R$ and $(y, x) \in R$ follows $x = y$. So the symmetric part of the relation is a subset (not sub-relation!) of the diagonal relation. Note that antisymmetry does not imply reflexivity. Some pairs (x, x) might be included, but not necessarily all. But we will not have any pair (x, y) where also (y, x) belongs to R and x and y are different.

A further important property of relations is transitivity. A binary relation R is transitive if for any triple (x, y, z) of domain elements (including equal elements) from $(x, y) \in R$ and $(y, z) \in R$ it follows that also $(x, z) \in R$. Actually, this property unifies two completely different aspects: the transitivity of equivalence and the transitivity of preference.

A nice aspect of transitivity is that we only need to show it for triples, and it automatically follows for longer chains of related elements. For example, if we have a quadruple (x, y, z, u) and $(x, y) \in R$, $(y, z) \in R$ and $(z, u) \in R$ and R is (triple-)transitive then it follows that $(x, z) \in R$ (from the triple (x, y, z)) and then that $(x, z) \in R$ (from the triple (x, z, u)). In the same follows, for any chain of related elements, the relativeness of its first and last elements.

Transitivity is also the heart of two combined properties of relations that play an important role in mathematics. One is the concept of an *equivalence relation.* This is any relation that is at the same time reflexive, symmetric, and transitive. An equivalence relation allows to split the domain into so-called equivalence classes. Elements in one such class are all in relation to each other, but no pair of elements from different classes belongs to R. Thus, these classes are pairwise disjunct, and their union is the whole domain. For some reasons that we will not detail here, this representation of the domain as set of equivalence classes of an equivalence relation R is sometimes written in quotient notation A/R.

The other is the concept of an order relation. Actually, it comes in two flavors. A relation that is reflexive, antisymmetric, and transitive is called a partial order (flavor 1). A domain with a superposed partial order then becomes a partially ordered set, or *poset* for short. Such relations refer to a "≥"-notation (more correctly, the real numbers along with the standard greater-than-or-equal-to relation). The corresponding ">"-notation then would be called a strict partial order (where the former than can be called a weak partial order). It is never reflexive but fully transitive. Since $x > y$ and $y > x$—by transitivity—would imply $x > x$ (where we use the >-symbol for any such relation) and thus it would contain a pair (x, x). We just excluded this a moment ago. So for a strict order relation (the word "partial" can usually be skipped), including the pair (x, y) it will never include the pair (y, x) as well—the relation is asymmetric.

When we skip the antisymmetry requirement for a partial order, we get a so-called pre-order. The most common appearance of such an order is if we compare domain elements by function values. So we have a mapping $f : A \to R^1$ (R^1 here means the set of real numbers, or any other poset) and we state that two domain elements x and y are in relation if $f(x) \geq f(y)$. So the elements of A are ordered according to their function values. Such a relation cannot distinguish between domain elements that map onto the same real number, thus the name "pre-order." To close the cycle: if we have a pre-order relation R_{po} over domain A, we define a new relation R_{pod} between domain elements. There the pair $(x, y) \in R_{pod}$ whenever $(x, y) \in R_{po}$ and $(y, x) \in R_{po}$. In other words, we take the symmetric part of a pre-order relation. This relation is an equivalence relation. If we consider the mapping approach to define a pre-order, its equivalence classes are domain elements that are all mapped onto the same real number.

If a relation is not transitive, then it means that we can find a triple (x, y, z) where x is in relation to y and y to z but not x in relation to z. But if this is the case, how about (z, x)? If this pair is not in relation, then —in some circumstances—the missing relation can be implied. Consider a path in a directed graph. If we have an arc linking nodes A and B and another arc linking nodes B and C, then there is a linking path from A to C, no matter if there is also an arc from A to C or not. However, if (z, x) also belongs to R, then we have a problem. It's called cyclicity of the relation.

More general: a chain of k domain elements $x_1, x_2, \ldots, x_k$ (where $k \geq 3$) such that for $1 \leq i \leq (k-1)$ we have $(x_i, x_{i+1}) \in R$ and also $(x_k, x_1) \in R$ then this is called a cycle of the relation, and the relation itself cyclic. A relation without any such cycle for any possible $k \geq 3$—like its directed graph counterpart—is called cycle-free.

Showing cycle-freeness of a relation can be a tedious task, since it has to be shown for any possible k. Contrary to transitivity, there is no way to deduce the non-existence of, e.g., 4-cycles from the non-existence of 3-cycles.

This definition of cycle-freeness also makes a problem with regard to the symmetric part of a relation. For example, we take the real-valued "$\geq$"-relation, and voila we easily find a 3-cycle: $(2, 2, 2)$ since in this triple (yes, we never excluded equal domain elements in chains) the first element is at least as large as the second, the second at least as large as the third, and the third is at least as large as the first. It is formally correct, but obviously not our intention to have "true" cycles. So there is a "trick" (e.g., suggested in [35]) to refer to the asymmetric part of R instead of R. It is also called a quasi-cycle if we have a chain of k domain elements where $(x_i, x_{i+1}) \in P(R)$ for $i = 1, \ldots, (k-1)$ and $(x_k, x_1) \in P(R)$. At least it helps to resolve the above ambiguity of repeated elements in chains.

A last comment: it was demanded that cycles should have at least 3 domain elements. How about $k = 2$ or even $k = 1$. In fact, a 1-cycle would be the same as non-reflexivity, where the domain element x under question is at the same time at the end and at the beginning of a chain. So such a relation must not contain any pair (x, x) for $x \in A$. This is also sometimes called anti-reflexivity.

For $k = 2$ we "read" the 2-cycle as a relation that contains at least one pair (x, y) such that $(x, y) \in R$ and $(y, x) \in R$. Thus, the non-existence of 2-cycles is the same as the asymmetry of the relation. This strengthens our consideration to focus on the asymmetric part of a relation when making corresponding definitions.

Once we have a relation, there are operators that allow us to produce a new relation. We have already seen two of them: the symmetric part and the asymmetric part are such operators. Now we give a few more of such operators.

1. The *complement* of a relation R over domain A is the set of all pairs (x, y) of domain elements that do not belong to R. Using the notation $\bar{R}$, we can write $\bar{R} = A \times A \setminus R$. Or in "point-wise" notation as

$$\bar{R} = \{(x, y) \in A \times A \mid (x, y) \notin R\} \tag{2.3}$$

2. The *converse* relation (or transpose relation) R^C results from swapping the elements in all pairs of the relation R as

$$R^C = \{(x, y) \in A \times A \mid (y, x) \in R\} \tag{2.4}$$

3. Combining both we get the *dual* relation R^d of R: $R^d = \bar{R}^C = \bar{R^C}$.

It might not be immediately obvious that the complement of the converse relation is the same as the converse of the complement. To see this, we have to consider when a pair (x, y) of domain elements belongs to $\bar{R}^C$.

1. The pair (x, y) belongs to $\bar{R}^C$ whenever its converse (y, x) belongs to $\bar{R}$.
2. The pair (y, x) belongs to $\bar{R}$ whenever $(y, x) \notin R$.

Now for the other case $\overline{R^C}$.

1. A pair (x, y) belongs to $\overline{R^C}$ whenever $(x, y) \notin R^C$.
2. A pair (x, y) is not element of R^C whenever $(y, x) \notin R$.

Comparing both point-wise specifications, we see that it's really the same relation. Thus, it is reasonable to define a dual relation R^d as the converse complement or the complement of the converse relation. We repeat the result here: a pair (x, y) of domain elements belongs to the dual relation of R if and only if the pair (y, x) does not belong to R.

Now we have already seen a lot of abstract specifications. Maybe it's time to show their interplay to prove a rather simple theorem about dual relations.

Theorem 2.1 *The asymmetric part $P(R)$ of a relation R over domain A is the intersection of R and its dual relation R^d.*

Proof Let us have a look at the specification of the asymmetric part $P(R)$ of some relation R over the domain A.

$$P(R) = \{(x, y) \in A \times A \mid (x, y) \in R \wedge (y, x) \notin R\} \tag{2.5}$$

However, the dual relation R^d—as the converse complement—is composed of all pairs (x, y) where $(y, x) \notin R$. Thus, the condition for the asymmetric part can also be written as

$$P(R) = \{(x, y) \in A \times A \mid (x, y) \in R \wedge (x, y) \in R^d\} \tag{2.6}$$

and the asymmetric part appears as intersection $P(R) = R \cup R^d$.

This has two immediate consequences that can be easily seen.

1. A relation R and its dual R^d have the same asymmetric part (follows from $(R^d)^d = R$).
2. If $R \subseteq R^d$ then $P(R) = R = P(R^d)$ (since then $R \cup R^d = R$).

This was only a small self-exercise, to see how proofs about relations and their properties are going on—at least if using the "point-wise" specification. It has to be said that such proofs are not very elegant, but usually the intricate matters of relations do not give us a better choice.

We will conclude our small journey into exploring relational basics by stating a way to create a new relation based on *two* relations. It is usually called a relational product given two relations R and S over the same domain A. It is very important that the domain is the same for both relations. Now (the constructive procedure comes first) we "concatenate" the relations such that we pick an element x from A, see to which elements from A it is in relation R. Let us call this set x_R. Then we pick an element y from x_R and see to which elements from A it is in relation S. Doing this for all domain elements, taking the role of x and all elements of the sets x_R, we get a number of sets. Their grand unification is the product of the relations R and S. Since it reminds more of sequencing the relations than "multiplying" them, a common notation is the semicolon, $R; S$ in our case. The actual definition of the relational product gives the condition when a pair belongs to the product relation.

Definition 2.11 Given two relations R and S over the same domain A. A pair (x, y) belongs to the *relational product* $R; S$ (or product relation, relational composition) if and only if there is a $m \in A$ such that $(x, m) \in R$ and $(m, y) \in S$.

It seems better to give a small example: consider a domain $\{a, b, c, d\}$ and the relations $R = \{(a, b), (a, c), (c, d)\}$ and $S = \{(b, d), (c, a)\}$. What is their product? Blindly following the definition would require us to check for all pairs (x, y) if we can find at least one "mediating" element m such that the first element x of (x, y) is in relation R to m and m itself is in relation S to y. We say mediating because its nature, value, or whatever else does not matter. It has to be there during the time of our checking and can vanish afterward. But such a test would be expensive, so we'd better follow the constructive principle given first. We see that a is in relation R to two elements of the domain, b and c. Among them, b is in relation S to d and c is in relation S to a. So, from checking the first domain element a it is in product relation $R; S$ to d and to itself. We have found two pairs. But we will not find any more. The domain element b does not appear on the left of any pair constituting the relation R, as well as d. The element c appears once, being in relation R to d, but then, d is not in relation S to any domain element. The meager outcome of our analysis is the relation $R; S = \{(a, a), (a, d)\}$. But we could get the result faster than checking 16 possible pairs taken from a domain of four elements, where each check considers four possible candidates for intermediate domain elements.

From this easy example, we can already see two "non-properties" of the relational product. If we look at what is $S; R$, we will find that its the set of pairs $\{(c, b), (c, c)$ and thus different from $R; S$. The relational product is not commutative. Moreover, $R; S$ neither contains R nor S—in fact, in our example, these sets are even disjunct and have no element in common. It shows that some care has to be taken to consider it as a composition operation in the classical sense.

There is much more to say about relational products. Based on it, other relational properties can be defined, procedures specified, and they are at the foundation of a mathematical discipline called relational algebra. We will show the usefulness of this operation by one example construction, the so-called transitive closure.

We mentioned already that a relation can appear like a transitive relation if we consider "transport processes." One can travel to the city of Oggersheim from Osaka even if there is no direct flight connecting both cities. We can go somewhere nearby and then take a train, bus, rental car, whatever. We just need to find that "hub." In the same sense, the travel network is transitive, which means that if we can go from A to B and also from B to C (probably changing the means of traveling) then we can consider A and C as linked. Trouble comes only if we cannot find such a place B at all. So can we see the aspect of the relational product somehow hidden here? Yes.

And its even better since we can consider multi-hop travel routes (to use a popular term from networking architectures) in the same way. So, having a travel network, how can we find what is connected to what at the end, allowing an unlimited number of hops (not thinking about the time it would need to pass all the security controls...)? We can use the relational product of the "travel connectivity relation" any number of times with itself. So at first, we have the places directly linked. Next, we have the places linked by one hub. Next, we take the places that are linked via two hubs—a few of them might be places already linked via one hub or directly, so our connectivity pattern might contain redundancies. And so on, ending up with a number of iterated applications of the relational self-product, where everything is linked to each other that can be linked via hubs. Since a travel network is finite, the process must stop. At the end, we cannot have more hubs than places to connect without getting some kind of cycle. Then we take the union of all these hub-linked scenarios for each number of iterations. Formally, it looks like this: we have a relation R over some domain A. With R^i where i is a natural number we denote the i-times iterated application of the relational product with itself. To avoid any misunderstanding: $R^1 = R$ and $R^{i+1} = R^i; R$. Then we construct

$$R^T = \Cup_{i=1,\ldots,\infty} R^i \tag{2.7}$$

and have a transitive relation R^T. But it's not only a transitive relation, it is in some sense the smallest transitive relation containing R. Any other transitive relation that contains R will have this relation R^T as a subset. Therefore, the special name is transitive closure.

2.2 Case Study: Pareto Dominance

In the foregoing section, we saw a number of abstract definitions and concepts. It's time to see them in practice, and we take the one relation that we already know a little (see Definition 2.5): Pareto dominance Relation. Let us explore it more.

First, have a look at the representation of Pareto dominance. In Definition 2.5 it was given in predicate representation. We used the logical truth value of a predicate in order to decide whether two vectors are in that Pareto dominance relation or not. It is the case when every component of a vector x is at least as large as the corresponding component of the other vector y, and, not to forget, at least one component is truly larger. So if we compare vector $x = (3, 4, 5)$ and vector $y = (2, 4, 4)$ than we see that $x_1 > y_1$, $x_2 = y_2$ and $x_3 > y_3$ so it neatly fits into our scheme: all components are at least as large and at least one (take index 1 or index 3) is larger. Thus $x >_p y$.

The relevant aspect of a predicate representation of a relation is that it works independently of the specific domain, as long as the predicate is properly specified. It makes one easily forget that there is a domain at all. So again, care has to be taken.

The functional representation gives for some vector x the set of all vectors y (from the domain) for which $x >_p y$. In general, any x dominates some hyper-volume, or better hyper-box, which is bounded by: (1) the coordinate (hyper)planes, (2) hyperplanes parallel to the coordinate planes that contain x. Or: we shift the coordinate hyperplanes until they hit the point x. Of course, this hyper-box has to be intersected with the domain. It can be seen that the hyper-volume dominated by some real-valued vector has a rather simple shape. This makes Pareto dominance also a convenient and imaginable relation.

There is nothing of interest to consider Pareto dominance as a directed graph or binary matrix. But we may note that Pareto dominance is the most natural (or: canonical) way of extending the standard larger relation between real numbers. In the case of one-dimensional vectors, both definitions coincide.

Then we continue to basic properties. Per definition, it is not reflexive. By the addendum "and at least one component has to be larger" we made sure that for any pair of vectors (x, y) that belong to the Pareto dominance relation it is $x \neq y$.

It is also not a complete relation. We can easily find pairs (x, y) where neither $x >_P y$ nor $x >_P y$. Such pairs of vectors are incomparable by Pareto dominance. For example: $x = (2, 3)$ and $y = (3, 1)$. Actually, having at least one smaller component is sufficient to break Pareto dominance. For two vectors having some components that are larger, and some others that are smaller, is already enough to make them incomparable.

We can see that it is an asymmetric relation. From $x >_P y$ follows for all component indices i that $x_i \geq y_i$. If we had $y >_P x$ at the same time, then also: for all component indices i it is $y_i \geq x_i$. Putting both together: $x_i \geq y_i \geq x_i$, which is another way of writing $x_i = y_i$. But we still have to consider the "addendum" by which at least one component of x has to be truly larger than the corresponding component of y. This is not possible if all components are equal. Ergo: from $x >_P y$ follows that $y \ngtr_P x$.

Let us modify Pareto dominance a little by adding the diagonal relation, i.e., all pairs (x, x) with x elements of the corresponding domain. Then we get a relation that is often called weak Pareto dominance. Its definition is a little bit shorter than Pareto dominance.

Definition 2.12 Given a domain $A \subseteq R^n$. Two vectors $x, y \in A$ are in weak Pareto dominance relation (written as $x \geq_P y$) if for every component index i we have $x_i \geq y_i$.

We have just added the diagonal relation, so this relation is by definition reflexive. It is still not complete, but it is no longer asymmetric. It is an example of an anti-symmetric relation. We have just seen that for two vectors being mutually in such a relation, it would need $x_i = y_i$ for all $i = 1, \ldots, n$. This means from $x \geq_P y$ and $y \geq_P x$ follows $x = y$.

Pareto dominance and weak Pareto dominance are easily linked to each other: Pareto dominance is the asymmetric part of the weak Pareto dominance. When we strip off weak Pareto dominance of all its pairs (x, y) where also (y, x), by virtue of antisymmetry, these are just the pairs (all pairs) where $x = y$. So we get back to the Pareto dominance, where we added the diagonal relation to get the weak Pareto dominance. Actually, it would have been more easy to define weak Pareto dominance as Pareto dominance and leave everything else to the asymmetric part machinery (it is one as we will see very often in the following). But historically, the naming conventions have developed in the way that was now given, and it has to be accepted. Nevertheless, there are thousands of printed publications, again and again adding that small sub-sentence "and at least one has to be larger" or some semantic equivalent. How much printing ink and paper could have been saved over the course of decades?

Now we come to the more tricky properties transitivity and cycle-freeness. In fact, Pareto dominance is transitive. Take three domain elements x, y, and z and assume $x >_p y$ and $y >_p z$. It means two things: (1) for all component indices i $x_i \geq y_i$ and $y_i \geq z_i$ and therefore $x_i \geq z_i$; (2) for some j, it must be $x_j > y_j \geq z_j$ and thus $x_j > z_j$. So both conditions are fulfilled: all components of x_i are at least as large as the corresponding components of z, and at least one is larger. It follows $x >_p z$. In the same way, argument (1) suffices to prove that weak Pareto dominance is transitive as well.

It is also a cycle-free relation. We recall that it means that for any "chain" of k domain elements $x_1, x_2, \ldots, x_k$ linked by the Pareto dominance sequence $x_1 >_p x_2 >_p \cdots >_p x_k$ it must never be $x_k >_p x_1$. We can use a simple trick to see this, referring to transitivity and cycle-freeness of the natural larger relation for real numbers. From $x >_p y$ it follows that the sum of components of x is larger than the sum of components of y. Remember: each component is at least as large as one truly larger than the component of y with the same index. If we denote the sum of components of a vector x as $h(x)$, then the above chain would imply $h(x_1) > h(x_2) > \cdots > h(x_k)$. But $x_k >_p x_1$ would imply $h(x_k) > h(x_1)$ in conflict with $h(x_1) > h(x_k)$. Another argument would be: transitivity of Pareto dominance gives that for the above chain $x_1 >_p x_k$. Asymmetry of Pareto dominance gives us that $x_k \not>_p x_1$ and we are fine. It's much shorter.

So to summarize: Pareto dominance relation is not reflexive, not complete, asymmetric, transitive, and cycle-free. Weak Pareto dominance is reflexive, antisymmetric, and transitive (and therefore a partial order), not complete, and (quasi-)cycle-free.

We have also mentioned ways to derive new relations from existing ones. In the case of Pareto dominance, it is interesting to see what is its dual relation. Since Pareto dominance is given in predicate representation, we need to manipulate the logical expression in order to find out about its dual. Remember that the dual relation is composed of all pairs $(x, y) \in A \times A$ where (y, x) belongs to the complement of R_p (this should indicate the Pareto dominance as a set). Pareto dominance is a set of pairs

$$R_p = \{(x, y) \in A \times A \mid (\forall i : x_i \geq y_i) \wedge (\exists j : x_j > y_j)\} \tag{2.8}$$

so we have to do two things: (1) swap the roles of x and y to get the converse relation; (2) negate the resulting logical expression to get the complement of the converse, i.e., the dual. Step (1) gives the expression

$$R_p^C = \{(x, y) \in A \times A \mid (\forall i : y_i \geq x_i) \wedge (\exists j : y_j > x_j)\} \tag{2.9}$$

We continue in a few steps as follows.

1. We use two sub-predicates $P_1 = (\forall i : y_i \geq x_i)$ and $P_2 = (\exists j : y_j > x_j)$ to simplify notation. The condition for (x, y) to belong to the converse becomes $P_1 \wedge P_2$, and its negation is the condition $\neg P_1 \vee \neg P_2$.
2. For the first part of the latter conjunction: $\neg P_1$ means $\neg(\forall i : y_i \geq x_i)$ or $(\exists i : y_i < x_i)$. So there is at least one component of y that is truly smaller than the corresponding component of x.
3. For the second part: $\neg(\exists j : y_j > x_j)$ or $(\forall j : y_j \leq x_j)$. It means all components of y are at most the corresponding components of x.
4. For the conjunction $\neg P_1 \vee \neg P_2$ to become true, either $\neg P_1$ is true, and if not $\neg P_2$ must be true.
5. So, combining with the foregoing results: either there is at least one smaller component in y (compared to the x-component at the same index) or for all indices i we have $y_i \leq x_i$ (now we can safely replace the index symbol j by i).
6. The case that there is no smaller component in y means that for all component indices i it is $y_i \geq x_i$. But this is the case where the first part of the conjunction $\neg P_1 \vee \neg P_2$ is not true, so the condition can only become logically true if the second condition becomes true. And we have just seen that this means for all these indices i we also have $y_i \leq x_i$. What follows is that then for all component indices i it must be $x_i = y_i$ and in total $x = y$.
7. So to summarize: either there is at least one component of y that is truly smaller than the corresponding component of x or both vectors are equal.
8. Changing to "x-language": either $x = y$ or there is at least one index component i such that $x_i > y_i$.

Congratulations, we could logically identify the dual relation of Pareto dominance! But we are not done yet. A moment of reasoning gives that it means the same as: if there is a component index i such that $y_i > x_i$ there must be an(other) index j such that $x_j > y_j$. What if there would be no such j? It would mean for all j $x_j \leq y_j$ and for at least one index i $x_i < y_i$. So it's neither $x = y$ nor is there any index i with $y_i < x_i$ in contradiction to our specification of the Pareto dominance dual relation. On the other hand, if no y-component is larger than the corresponding x-component (so we do not have to see if there is another component being smaller) than it also means that either $x = y$ or at least one y-component is truly smaller.

In summary, we can formulate the dual to Pareto dominance as follows: for any component index i with $y_i > x_i$, there is another component index j such that $x_j > y_j$. If the vector components, for example, represent allocated shares of goods to individuals, then it reads: for any individual being better off there is at least one individual being worse off. It is what we have met several times in the foregoing chapter and called Pareto efficiency there. For the files, let us fix it by definition.

Definition 2.13 (*Pareto Efficiency*) Given a domain $A \subseteq R^n$ and two elements x and y of A. It is said that x is *Pareto efficient* against y (written as $x \geq_e y$) if and only if for any component index i with $y_i > x_i$ there is at least one component index j with $x_j > y_j$.

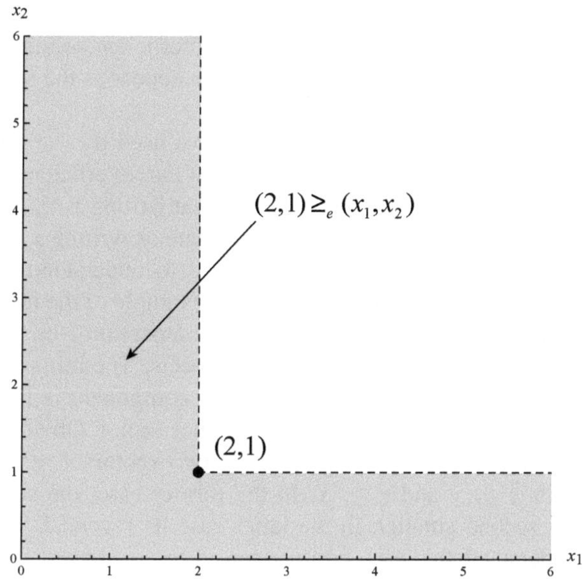

Fig. 2.1 Set of points to which (2, 1) is in Pareto efficiency relation (the domain here is the positive quadrant)

As an alternative representation, Fig. 2.1 shows, as an example, the area of vectors (or better: points) with non-negative components to which (2, 1) is in Pareto efficiency relation. We yield the general shape of the area to which a point x is in Pareto efficiency relation by shifting the coordinate (hyper)planes in parallel to themselves until they hit the point, and then take the union of all such enclosed segments—excluding the points on the shifted hyperplanes themselves. At the end, we add the point itself, and intersect with the domain, of course. It can be easily seen that Pareto dominance is a part of that related set.

The attentive reader will have noted that, while studying the rate allocation problem in the foregoing chapter, the term "efficient solutions" was used somewhat ambiguously. Sometimes we characterized it as a solution where any increase for one user must be accompanied by a decrease for some other user. The other time we said that there is no way to allocate at least as much as an efficient solution to all users, and more for at least one user. Now we can formalize the issue more clearly. In the former case, we spoke about the greatest element of the Pareto efficiency relation, in the latter case about a maximal element of the Pareto dominance relation (in other words: about a non-dominated solution). It appears to be the same.

The link goes even a little bit further. According to Definition 2.13 if all x_i are at least as large as the related y_i and at least one is larger, there is no index j such that $y_j > x_j$. So the predicate to check for the Pareto efficiency relation is fulfilled. It means that from Pareto dominance follows Pareto efficiency, or from $x >_p y$ follows $x \geq_e y$ and (over the same domain) the Pareto dominance relation is a subset of the Pareto efficiency relation, i.e., its dual relation. Now remember our little Theorem 2.1 and the following comment? It follows that the asymmetric part of the Pareto efficiency relation is the Pareto dominance relation. So Pareto efficient allocations appear as greatest Pareto efficient solutions, as well as maximal Pareto efficient solutions. The fact that both appear as the same set of solutions will be our concern in the next section.

One may have noted that we have used the "$\geq$"-notation for Pareto efficiency. Of course, we have to appreciate that Pareto efficiency is not asymmetric. And from the foregoing paragraph, we know that writing $x >_e y$ (with subscript "e" indicating Pareto efficiency) amounts to the same as writing $x >_p y$ since both have the same asymmetric part. But it seems hard to understand Pareto efficiency as a way of comparing two vectors. It is a good example of the fact that it is only the asymmetric part of a relation that represents comparison—or, with some slight abuse of the English language "betterness"—aspects. The remaining (symmetric) part includes pairs of vectors where at least one component is larger and one smaller than the corresponding component of the other vector. Obviously, this is a mutual relation.

Take as a simple example, the two vectors $x = (2, 3)$ and $y = (3, 2)$. We have both $x \geq_e y$ and $y \geq_e x$. In the former case, the second component is larger and the second smaller, in the latter case its reversed. The best thing we might try to understand this as a comparison is to consider it as an "unacceptable transition" from x to y. It means we are safely in a state x and think about changing to y. What will happen? Component 1 is a clear winner of this transition since $y_1 = 3 > x_1 = 2$. But component 2 is objecting, since it will become smaller at the same time. So we would not accept that transition. Then a Pareto efficient solution is a solution where no transition to any other solution is acceptable.

To conclude this discussion (that might appear petty to someone) we will show that this Pareto efficiency relation might also be a record holder in *missing* relational properties.

1. It is not antisymmetric. We have just seen the example $x = (2, 3)$ and $y = (3, 2)$ where both $(x, y) \in R_e$ and $(y, x) \in R_e$ but it does not follow $x = y$. It is not asymmetric for the same reason.
2. It is not symmetric as well. It includes Pareto dominance, which is asymmetric.
3. It is not transitive. We take a triple of two-dimensional vectors $x = (3, 3)$, $y = (2, 5)$, $z = (4, 4)$. From x to y one component gets larger and the other smaller, from y to z one component gets larger and the other smaller, but for $z = (4, 4)$ both components are larger than for $x = (3, 3)$ and so we have an example where $x \geq_e y$ and $y \geq_e z$ but not $x \geq_e z$.
4. The same example appears as a circular relationship, since $x \geq_e y \geq_e z \geq_e x$. But it does not mean that Pareto efficiency is cyclic since we wanted to refer to the asymmetric part when speaking about cycle-freeness. The asymmetric part is the Pareto dominance, and this is a cycle-free relation. So Pareto efficiency is cycle-free, even if it's full of triples (and longer chains) of circularly related elements.

Maybe now we have seen enough about Pareto efficiency. We will conclude our case study with the consideration of a relational product. We might look for the relational product of Pareto dominance with itself. We will do this since it is a good example for the relevance of the domain of a relation for such products. Let us consider the definition of the relational product here (using R_p as a notation for Pareto dominance).

$$(x, y) \in R_p; R_p \longleftrightarrow \exists m \in A : (x, m) \in R_p \wedge (m.y) \in R_p \tag{2.10}$$

By transitivity of Pareto dominance, from $(x, m) \in R_p$ and $(m, y) \in R_p$ follows $(x, y) \in R_p$, and therefore whenever a pair (x, y) belongs to the relational product of Pareto dominance with itself, then it also belongs to Pareto dominance. So $R_p^2 \subseteq R_p$. But not $R_p^2 = R_p$! This is only true for convex (continuous) domains. Equality of R_p^2 and R_p would mean that for any $(x, y) \in R_p$ we can find at least one "intermediate" m such that $(x, m) \in R_p$ and $(m, y) \in R_p$. So we have to make sure that for every pair (x, y) where x Pareto- dominates y we can indeed find a third vector m who links both elements such that they can join the relational product. If the domain is convex, we can take any convex combination between x and y as such an m, for example, the average $m_i = (x_i + y_i)/2$ for all component indices i. Obviously from $x_i \geq y_i$ follows $x_i \geq m_i \geq y_i$ and from $x_j > y_j$ for some j follows $x_j > m_j > y_j$.

But if the domain is not convex, the relational product might differ from Pareto dominance. Just consider a discrete domain, like $\{(2, 2), (3, 3), (4, 4)\}$. The pair $((3, 3), (2, 2))$ belongs to the Pareto dominance relation over this domain, but it will not be part of the relational product of Pareto dominance over this domain with itself. The relational product, in fact, will only contain the pair $((4, 4), (2, 2))$ where we have the intermediate vector $(3, 3)$, and nothing else. In general, for discrete domains, it appears hard to predict the relational product.

2.3 Why the Asymmetric Part

Coming back to optimality represented by relations and the definitions for greatest and maximum elements. The first question one might have is why we have one definition (greatest element) directly referring to the relation, while the other definition (maximal element) refers to the asymmetric part.

There are at least two reasons. The one is very simple: without this adjustment, the definition of the maximum set would be just a spoiler. Before we continue, let us see two definitions.

Definition 2.14 (*Vector Relation*) A *vector relation* is a binary relation over a domain $A \subseteq R^n$.

All relations that we have listed so far are vector relations. The main reason to speak about "vectors" here is to refer to domain elements that are constituted as an indexed set of components. We could have used the word "point" as well (and we have already done so). By $\Pi(x)$ we denote a rearrangement of the components, as for example the transition from vector $(1, 3, 4)$ to vector $(3, 4, 1)$.

Definition 2.15 (*CUP Relation*) A vector relation R over domain A is closed-under-permutation (CUP) if from $(x, y) \in R$ and $\Pi_1(x) \in A$ as well as $\Pi_2(y) \in A$ follows that $(\Pi_1(x), \Pi_2(y)) \in R$.

Pareto dominance is not a CUP relation. We can take the related pair $x = (10, 5, 7)$ and $y = (8, 4, 6)$. But a permutation of x-components like $(5, 10, 7)$ will not dominate y (seen as the identical permutation of y). We will see CUP relations soon, but for now, a simple example is to compare two vectors by the sum of their components. These sums will not depend on the order of the addends. If a CUP relation is also reflexive, i.e., each domain element is in relation to itself it means that each domain element is also in relation to any permuted version of itself that still belongs to the domain.

Now, coming back to the definition of maximal elements. Assume we would not use $P(R)$ but R itself. Then what happens if it is a relation where all domain elements are in relation to themselves? So there would not be any maximal element since for each element, there exists at least one domain element which is in relation to it: the element itself. The asymmetric part strips off all these mutual relation pairs.

There might be the "patch" by saying that x' is maximal whenever there is no $x \neq x'$ for which $(x, x') \in R$. In fact, this definition can be found here and there. But it will not help much to save the day. Our CUP relations come into play. If we did not use $P(R)$ in the definition of the maximum set here then also in most cases these relations would never have maximal elements since there are always permuted versions of themselves (at least one should usually belong to the domain) that are related to them.

Probably, we might still have some advocates for the solutions to ban permuted versions as well. But then we get into epicycles: take now scalar relations, i.e., relations where we compare vectors by comparing function values of their components. In the simplest case, it is the sum, but how many ways are there to achieve the same sum from components of totally different and unrelated vectors? A vector like $(10, 2, 5)$ could never be maximal since the vector $(8, 7, 2)$—if it also belongs to the domain (if not, we find another one)—that has the same sum and is thus in this greater-or-equal-sum relation to $(10, 2, 5)$. If someone argues now that we should use greater-sum instead of greater-or-equal-sum, than we have finally reached our goal to convince that it's better to use the asymmetric part. If doing so, all these nasty issues are resolved.

This is one reason, made for the practitioner (and a little bit of reference to Occam's razor). On the other hand, we do not have such problems with greatest elements. The only trivial obstacle here is that being in relation to all domain elements implies being in relation to itself. So a relation must be reflexive in order to have the greatest elements. But in practice, this can be easily achieved by using a substitute relation that is the same relation unified with the diagonal relation. For example, we can add all domain pairs (x, x) to Pareto dominance—and we already know this relation is called weak Pareto dominance. An asymmetric relation cannot have the greatest elements. Otherwise, there is no problem with CUP relations or mapped comparisons mapping to the same value.

We may also remind on the appearance of the asymmetric part for specifying cycles of a relation. The reasons for using the asymmetric part appear quite similar here. But another issue is that each relation is a kind of hybrid between expressing betterness and indifference. This is the meaning of the fundamental decomposition

$$R = P(R) \cup I(R) \tag{2.11}$$

where $P(R)$ represents the specific way of a relation to compare two domain elements with an ordered outcome, while $I(R)$ represents the specific way of disambiguating two domain elements. We have many related terms for both: preference, coming before, causing, being better, larger, smaller, tastier, cheaper, faster, or any other use of the grammatical comparative for the asymmetric part, and equal, similar, nearby, indifferent, identical for the symmetric part. How it comes that each binary relation shares both aspects is a different question that can hardly be answered here.

Now if we have a greatest domain element and there are other domain elements in a symmetric part relation to it then these other elements can be understood of being greatest as well. Thus, this would not be the case for maximal elements, which should not undergo mutual relations with each other.

We hope that these explanations help to satisfy the critic. But then a second question might come up: why do we have two flavors of extremality at all? What is the difference between the greatest and maximal elements? This is a deeper question and might not have an ultimate answer. Here we will only provide some arguments that might help to understand the distinction.

At first, both concepts are not totally independent. Let us take a complete partial order relation. Maybe again we take the simple example to compare two vectors by the sum of their components. A vector with the largest sum is supposed to be the "optimum"—but is it the greatest element or the maximum element? The relation here is to say that for two vectors x and y from some suitable domain it is $x \geq_s y$ if and only if $h(x) \geq h(y)$ (note that we have used the "$\geq$"-notation, and remember that we used the notation $h(x)$ for the sum of components already before). Then, for the greatest vector x', it must be $h(x')$ greater than or equal to any other h-value of a domain element. So it's one vector (there can be multiple) with the largest sum of components. But for a maximum vector, we need first to know the asymmetric part of the sum-comparison relation. This is the relation $x >_s y$ iff $h(x) > h(y)$. Then, maximality of x' means there is no other domain element x with $h(x) > h(x')$. Also here, x' must be a vector with the largest sum of components.

So in this "scalar case" there is actually no difference between greatestness and maximality. We like to ask which specific properties of a relation make this happen. It is time for a theorem.

Theorem 2.2 *If a relation is complete, then its maximum set and best set are equal.*

Proof Assume a complete relation R over the domain A where the maximum set (of all maximal elements) and best set (of all greatest elements) differ. Assume first an $x \in A$ that is maximal but not greatest. The fact that x is not the greatest element means that there must be an element $y \in A$ where $(x, y) \notin R$. Since R is a complete relation, it follows $(y, x) \in R$. The pair (y, x) does not belong to the symmetric part since $(x, y) \notin R$—this was the starting point of our chain of reasoning. So we would have a pair $(y, x) \in P(R)$ in contradiction to the assumption that x is a maximal element. Now the other case: assume x is greatest but not maximal. Not maximal implies that there is an element $y \in A$ such that $(y, x) \in P(R)$. Then, by the specification of the asymmetric part, it means $(x, y) \notin R$. So x would not be the greatest element.

If a relation is not complete, the maximum set and best set can differ or not. Take for example Pareto dominance, which is not a complete relation. We use two-dimensional vectors and as a domain the unit square (including its boundary). Formally, the domain is composed of all points (x, y) where $0 \leq x, y \leq 1$. Then the point $(1, 1)$ is both maximal and greatest, and there is no other extreme point. On the other hand, if we use the domain of two-dimensional points specified by $x + y = 1$ and $0 \leq x, y$ then all domain points are maximal but none is greatest. It follows from the fact that no pair of domain elements is in a Pareto dominance relation to each other at all. Since for both points their two components sum up to 1 one component must be larger and the other smaller than the corresponding component of the other point.

So the first thing to fix is that there is no difference between maximality and greatestness for the most popular case of using a scalar comparison function for specifying the relation (be it an aggregation function, cost function, fitness function, quality function etc.). We will face a difference only when starting to use “true” pairwise comparisons—that cannot be replaced by a comparison of scalar values (they would be complete relations otherwise)—like Pareto dominance.

The next theorem states another redundancy between both concepts of extremality.

Theorem 2.3 *The greatest elements of a relation R are maximal elements of its dual relation R^d.*

Proof It suffices to rewrite the definition of a greatest element. It is an element x where for all domain elements $(x, y) \in R$ holds, thus $(y, x) \in R^c$ and $(y, x) \notin \bar{R^c} = R^d$. Since $P(R^d) \subseteq R^d$, it also follows that for all y of the domain $(y, x) \notin P(R^d)$, and x is a maximal element of the dual relation.

The converse is not true, maximal elements of a relation are not greatest elements of its dual relation. A short example can show this. Take the domain $\{a, b, c\}$ and the relation $R = \{(a, b), (b, a), (b, c)\}$. The complementary relation is $\bar{R} = \{(a, a), (a, c), (b, b), (c, a), (c, b), (c, c)\}$ and its converse is $R^d = \{(a, a), (c, a), (b, b), (a, c), (b, c), (c, c)\}$. We see that a and b are maximal elements of R since they do not appear in the second position of any pair in the asymmetric part $P(R) = \{(b, c)\}$. But both are not greatest elements of R^d: for a we miss the pair (a, b) and for b we miss the pair (b, a). It is not an accident that this happens for the pairs that belong to the symmetric part of the relation. If an element is maximal despite appearing in a pair (y, x) because also $(x, y) \in R$ and thus $(y, x) \notin P(R)$ then x and y will be incomparable in the dual relation: neither (x, y) nor (y, x) appear there.

We have already seen an example for the statement of Theorem 2.3: the Pareto efficiency. Its greatest elements are all maximal elements of the dual relation, Pareto dominance. The greatest elements of Pareto efficiency are vectors where a possible increase of one component must result in the decrease of another component in order to be feasible (i.e., for the result of the change to still belong to the domain). So the greatest elements are Pareto efficient solutions. They are maximal elements of Pareto dominance: we have no way to improve all components at once, including at least one truly.

Continuing with Pareto efficiency, we can refine the above theorem a little. We noted already that all maximal elements of Pareto dominance are greatest elements of Pareto efficiency as well. It is because Pareto dominance is asymmetric.

Theorem 2.4 *The maximal elements of an asymmetric relation R are the greatest elements of its dual relation R^d.*

Proof Also, here some rewriting of the definition of a maximal element x is sufficient. Since R is asymmetric, it is $P(R) = R$, and it follows from the maximality of x that there is no y of the domain such that $(y, x) \in P(R) = R$. Or: for all y $(y, x) \in \bar{R}$. Or: $(x, y) \in R^d$. So x is indeed a maximal element of R^d.

A small comment is necessary to avoid any confusion: we have seen that Pareto dominance is the asymmetric part of Pareto efficiency. So, maximality of Pareto efficiency is the same as maximality of Pareto dominance. At the same time, Pareto dominance is the dual relation of Pareto efficiency. But it does not mean that maximal Pareto efficient solutions, which are also maximal elements of Pareto dominance, are at the same time always the greatest elements of Pareto dominance. Since Pareto dominance always has maximal elements (we will discuss this in a moment), it would also always have greatest elements. Reality teaches us that this is, unfortunately, not the case. The conclusion is wrong because Pareto efficiency—contrary to Pareto dominance—is not asymmetric. So even if duality appears as a mutual link between two relations, the issues of extremality are "oriented"—in both directions, the greatest elements are maximal elements of the other. In case one relation is asymmetric, the other cannot be asymmetric as well, but in addition to the former greatest elements of the other are also former maximal elements of the latter. But vice versa it is not necessarily true. To put it short: maximality and greatestness are not really dual concepts.

One more theorem states that the greatest elements are special cases of maximal elements.

Theorem 2.5 *For a relation R over a domain A, all greatest elements are also maximal elements.*

Proof A greatest element x is such that for any $y \in A$ it is $(x, y) \in R$. Then we have a look at the pair (y, x). If it is also in R then the pair (x, y) belongs to the symmetric part and $(x, y) \notin P(R)$. If it is not in R then also $(y, x) \notin P(R)$ since $P(R)$ is a subset of R. In either case for all $y \in A$ we get $(y, x) \notin P(R)$ and x is maximal.

It means that the best set is always a subset of the maximum set. It covers the case that the best set is empty—which is rather likely and gives the idea of the maximum set as an extension of the best set. If there are no greatest elements available, we still have "good" elements via maximality. Needless to say, there are maximal elements that are not greatest elements. Just take the relation $R = \{(a, b), (a, c)\}$ over the domain $A = \{a, b, c\}$. Here, a is a maximal element. But the relation does not include (a, a) in order to make a also a greatest element.

We will conclude our examination of the interplay of maximal and greatest elements by looking for implied relations. What means "implied relations"? We have already noted that for relations, because they are also sets, set operations and properties apply directly. One is subsethood. If for a relation R_A we have another relation R_B (over the same domain) such that $R_A \subseteq R_B$ what could this mean for extreme elements?

Definition 2.16 (*Implied Relation*) If for two relations R_A and R_B over the same domain A it is $R_A \subseteq R_B$, then relation R_A implies relation R_B.

As an example, the Pareto dominance relation implies the larger-sum-of-components relation that we have already used several times as an example. But be careful about the equality case. If $x >_p y$ then their sums of components are guaranteed to be in relation $h(x) > h(y)$ and not only $h(x) \geq h(y)$. So it is the asymmetric relation Pareto dominance that implies the asymmetric part of the larger-sum-of-components relation (no matter how we "interpret" the case of equality). We can make an important statement for this case.

Theorem 2.6 *If for two relations R_A and R_B over the same domain A, the asymmetric part of R_A implies the asymmetric part of R_B, then the maximum set of R_B is a subset of the maximum set of R_A.*

Proof We consider a maximal element x of R_B. It means there is no $y \in A$ such that $(y, x) \in P(R_B)$. Since $P(R_A) \subseteq P(R_B)$, it also means that there is no y such that $(y, x) \in P(R_A)$ and x is a maximal element of R_A as well.

The use of the asymmetric part here is relevant. If we only know that the relation R_A implies R_B, then we cannot make a safe statement about the influence on the maximum sets. Consider two cases:

1. In R_A there is an element x to which only one pair (y, x) exists. Thus, the pair is in the asymmetric part, and x is not maximal. Now we extend the relation R_A by a single pair (y, x). In this moment (x, y) belongs to the symmetric part, and since no other domain element is in relation to x it becomes maximal. The size of the maximum set grows by one.
2. Another case: we have a maximal x of R_A and some y such that neither (x, y) nor (y, x) is in R_A. Now we extend R_A by adding the pair (y, x). Then (y, x) belongs to the asymmetric part of the extended relation, and x is no longer maximal. The size of the maximum set shrinks by one.

So there can be cases where the maximum sets grow or shrink by successively adding pairs of domain elements to the relation. What happens depends on the specific relation, and no general statement is possible. It is only safe to check whether the asymmetric part of one relation implies the asymmetric part of the other.

We have already made use of Theorem 2.6 in Chap. 1 by what we called "Pareto trick"—for some ways of specifying optimality, the restriction to the efficient solutions was possible: the maximal product, the majorants, etc. In all cases, it was rather easy to see that the relation condition is fulfilled for increasing components. For example, we compare two vectors by their product of components. If we compare a vector x this way with some vector x' that Pareto dominates x it follows that the product of components of x must be (truly) larger. So Pareto dominance implies a larger component product, and both are asymmetric relations. Therefore, a vector with a maximum product of components must belong to the maximum set of Pareto dominance. Actually, the issue about the implication of the asymmetric part—this is what is needed to ensure the statement about maximum sets—was not clearly considered in the examples before. Now, with the framework of extremality of binary relations, it appears more clearly what has to be shown. It is more elegant if we consider the (later on, we will introduce its name proportional fairness) "true" relations like: x is in relation to y (both of dimension n and with positive components) whenever

$$\sum_{i=1}^{n} \frac{y_i - x_i}{x_i} \leq 0 \tag{2.12}$$

It is an antisymmetric relation. We see for each component index i

$$\begin{aligned}
(x_i - y_i)^2 &\geq 0 \\
x_i^2 + y_i^2 &\geq 2x_i y_i \\
\frac{x_i}{y_i} + \frac{y_i}{x_i} &\geq 2 = \frac{x_i}{x_i} + \frac{y_i}{y_i} \\
\frac{x_i - y_i}{x_i} + \frac{y_i - x_i}{x_i} &\geq 0 \sum_{i=1}^{n} \frac{x_i - y_i}{x_i} + \sum_{i=1}^{n} \frac{y_i - x_i}{x_i} \qquad \geq 0
\end{aligned}$$

and the equality only holds for $x_i = y_i$. So if we would have a case where x is in that relation to y and vice versa, then the last sum must be smaller than or equal to 0 at the same time, and this is only possible if it's 0. Which implies $x_i = y_i$ for all component indices i: finally $x = y$. The relation is antisymmetric, and the asymmetric part appears by removal of the pairs (x, x). Now, if we have $x_i \geq y_i$ and at least one $x_j > y_j$, the inequality is directly fulfilled since each summand is 0 or negative and at least one is negative. So Pareto dominance implies our relation, and we have to look for its maximal elements within the maximum set of Pareto dominance, the efficient solutions.

But Theorem 2.6 is also some kind of bad news about maximum sets. It means that the more specific relation (one can also say more sparse relation) tends to have larger maximum sets. If we represent the relation by a predicate and we combine this predicate with another one by conjunction (i.e., logical AND) then the new relation will be a subset of the original one, and the maximum sets will tend to grow (we say "tend" because the influence on the asymmetric part depends on the specific predicates and this cannot be a general statement).

For the greatest elements, there is a similar statement, but it does not have much practical value.

Theorem 2.7 *If relation R_A implies relation R_B then the best set of R_A is a subset of the best set of R_B.*

No need for proof. If we go from R_B to R_A by successively removing related pairs, we can basically lose the greatest elements, but not yield new ones.

Speaking about the size of extreme sets, we want to devote the remaining part of this section to the question of whether extreme elements exist at all. In the following, we assume that the domain of a relation is discrete and finite. This is because for statements about continuous domains, the set-theoretic framework that we have used so far will not be sufficient, and we would need concepts from topology. For example, taking the standard larger relation of real numbers and the sequence $x_i = (2^i - 1)/2^i$ for $i = 1, 2, \ldots$ as the domain A. This series is converging to 1 from below, but there is no i such that $x_i = 2$. So in this domain we have neither a maximal nor a greatest element, for every x_i there is an index j such that $x_j > x_i$ (no maximal element) and $x_i \not\geq x_i$ (no greatest element). Of course, topology expands the set notion by defining a system of open neighborhoods where 1 is specified as a limit point of the series, or also as "supremum." But basically, we are looking for elements of the domain and may ignore topological properties of relations for the time being.

For finite domains, we have an important statement about maximum sets.

Theorem 2.8 *If a relation R over a finite and non-empty domain A is cycle-free (w.r.t. its asymmetric part), then its maximum is not empty.*

Proof We take the asymmetric part $P(R)$ and assume that there is no maximal element. It means that for each $x \in A$ we can find a d_x (should mean "dominator of x") where $(d_x, x) \in P(R)$. We pick one $x_1 \in A$ randomly. There must be a dominating element d_{x_1} different from x_1 in A with $(d_{x_1}, x_1) \in P(R)$. Take this dominating element as element $x_2 \neq x_1$ to start constructing a series. There must be another domain element x_3 that is different from x_1 and x_2 such that $(x_3, x_2) \in P(R)$. It is not possible that the only choice for x_3 would be x_1 or x_2 (it means except x_1 and x_2 for no other $y \in A$ it is $(y, x_2) \in P(R)$): x_2 is not possible because (x_2, x_2) cannot be in the asymmetric part of a relation, and also (x_1, x_2) because already $(x_2, x_1) \in P(R)$. Next we look for an x_4 different from x_1, x_2, x_3 and $(x_4, x_3) \in P(R)$. Also, this must exist. To be different from x_2 and x_3 again follows from membership in the asymmetric part. If then only x_1 would be possible, then the relation would have a cycle: $(x_2, x_1) \in P(R)$, $(x_3, x_2) \in P(R)$ and $(x_3, x_1) \in P(R)$. This way, we can continue until we exhaust the domain. For the last element in the series, no dominating element would remain, in contradiction to our assumption that there is no maximal element.

For example, we know that Pareto dominance is cycle-free. Thus, every finite non-empty set of vectors has a non-empty maximum set. At the same time, there might be no greatest element. Its safe to say that it is even rather unlikely to have the greatest elements. It means that our life is full of trade-offs.

We also have some conditions that guarantee the existence of greatest elements. According to [25]:

Theorem 2.9 *If a relation R is complete and transitive, each finite set S has at least one greatest element.*

Proof For each element $x \in S$ consider the count $C(x)$ of elements of S dominated by x. Then consider an x with a maximum value of this count. Assume there is an $x' \neq x$ not dominated by x. From completeness follows that x' dominates x. From transitivity follows that x' dominates all elements dominated by x. Thus $C(x') > C(x)$, a contradiction. Therefore, an x with maximal $C(x)$ dominates all elements of S.

Neither transitivity nor completeness alone ensures the existence of greatest elements. The relation $R_1 = \{(x, y), (y, z), (x, z), (u, v)\}$ is transitive but has no greatest element. The relation $R_2 = \{(x, y), (x, z), (y, z)\}$ over the domain $\{x, y, z\}$ is complete but also has no greatest element.

It follows another difference between maximum sets and best sets with regard to their evolution from empty relations to full relations. For the empty relation, which equals its asymmetric part, no element is dominated by any domain element. Thus, every domain element is maximal, but none is greatest. If we successively add pairs until we end up with the full relation (i.e., $A \times A$), then again every element is maximal. It's because the asymmetric part is empty: for any $(x, y) \in A \times A$, also $(y, x) \in A \times A$. But every element now is greatest. And the progress of the best set growth is rather simple: it will grow monotonically by size. The progress of the maximum set can be quite different.

We start from the empty relation and add some pair (x, y). Then y will drop out, and the size of the maximum set (we consider a finite domain) will be $N - 1$, where N is the number of domain elements. Next, we add (y, x). Now again the asymmetric part is empty and we are back to N maximal elements. Continuing this way by adding one pair, then its converse, the size of the maximum set will alternate between N and $N - 1$—thus staying nearly constant N over the progress to the full relation.

On the other hand, we first start to construct a cycle of all domain elements. It means we order the domain elements somehow as $A = \{x_1, x_2, \ldots, x_N\}$ then add the pair (x_2, x_1), followed by (x_3, x_2), (x_4, x_3) and so on. At step N we add (x_1, x_N). This relation is asymmetric, and it has a cycle including all domain elements. No element is maximal since for every element, there is another domain element related to it. The maximum set size goes from N to 0. We can maintain this cycle within the asymmetric part by adding unrelated pairs (i.e., pairs different from the N pairs (x_2, x_1), (x_3, x_2), ...) before we have to add these converse pairs at the last N steps and when the maximum set size grows back to N.

This shows two extreme cases, where for one, the transition from empty relation to full relation nearly all domain elements stay maximal, and for the other, nearly no domain element is maximal. However, in practice, we are quite interested in small but non-empty maximum sets (as we want to solicit and select optimal elements—and what kind of bad optimality would it be to have nearly all feasible solutions as optimal ones!) and thus it is an indication that proper relation design is an essential step of specifying optimality.

2.4 Other Ideas About Optimality and Best Choices

In the foregoing section, we have seen the standard way of using relations to select optimal elements from a domain. In general, it can be assumed that we can only select something good from a pool of solutions if we have at least a rough idea about comparing different solutions. However, the specific way of selecting is not uniquely determined from more basic principles. It seems most attractive to find greatest elements of a relation—but if they do not exist, then maximal elements are a good alternative.

We have missed an important aspect of optimization so far. Not for negligence, but just because it does not formally pose a big challenge. There are situations where we are not interested in something maximal, or "most" but minimal, or "least." The most typical case is having a cost function—costs seem to be something that should be as small as possible. Another related term is risk. And several others.

As much as we could define maximal and greatest elements, we can define minimal and least elements. Here are the definitions.

Definition 2.17 (*Minimal Element*) Given a relation R over the domain A. A *minimal element* of R is any x which is not in relation $P(R)$ to any $y \in A$. All minimal elements comprise the minimum set of R

$$Min(P) = \{x \mid \nexists y \in A : (x, y) \in P(R)\} \tag{2.13}$$

Definition 2.18 (*Least Element*) Given a relation R over the domain A. A least element of R is any x to which every $y \in A$ is in relation. At least one element comprises the worst set of R

$$Wor(P) = \{x \mid \forall y \in A : (y, x) \in R\} \tag{2.14}$$

Note that we are also using the asymmetric part for the minimal elements, for the same reasons as using the asymmetric part for specifying maximal elements. Most of the statements given about maximal and greatest elements can be transferred into statements about least and minimal elements. But we will not go into details here. But one point has to be considered more seriously. In practice, these definitions of minimality will not be of much interest. There are "intrinsic" ways to consider the opposite of a relation. These notions are of a formal nature and cannot be captured within a pure set-theoretic framework. For example, take the weak Pareto dominance relation. It is based on all $x_i \geq y_i$ for two vectors x and y and related component indices i. Now, if considering minimality, it appears natural to change this definition to a reversed version $x_i \leq y_i$ for all i. For this relation, we still have to consider maximality (maximal and greatest elements) according to the original definitions. However, for the moment, we have no problem, since minimal weak Pareto dominance or maximal elements of the reversed version coincide. But soon we will encounter cases where both are different, and even cases where such a reversed version cannot be easily provided. So, for now, we note that the issue of minimality will need further consideration.

But back to the maximality issues. We said before that some preference would be for the greatest elements. But in many circumstances, they do not exist. We can find another, more elaborate condition for the existence of greatest elements from the "King Chicken Theorems" [25], which does not need transitivity. The somewhat strange name "king chicken" does not only fiddles with our imagination of a "strong chicken" but also refers to a general concept that applies to any relation. It was observed in barns that if two chickens (of the same gender) meet, usually one chicken pecks the other in the neck. It indicates that the pecking chicken is somehow superior to the pecked one. So, between any pair of same-gender chickens, there is such a binary relation. But does it mean that there is a "boss chicken" that will peck all other chickens (it does not seem so relevant to consider reflexivity of such a relation as well)? Observations show that this is probably not the case. As we know, having a complete relation is not enough to ensure the greatest elements—and such a "boss chicken" would be exactly the greatest element of a pecking relation. In [25], the question was analyzed if there is a way—at least in theory—to ensure the existence of a somehow greatest chicken with regard to its pecking strength. The idea to solve this problem was the provision of a little bit smooth version of "greatestness." The *king chicken* was born: a king chicken Ch_A was defined as a chicken that, for any other chicken Ch_B, either pecks Ch_B directly or pecks at least one chicken Ch_C that pecks chicken Ch_B.

This definition might sound familiar. In fact, it is the same way as we have defined relational products. Formally, a king chicken appears as a chicken that is greatest with regard to either relation R directly, or $R; R = R^2$. The latter means exactly that $(x, y) \in R^2$ if there is at least one $m \in A$ such that (x, m) and (m, y) both belong to R. Now, under the weaker condition that R is a complete relation, the following—maybe a little bit surprising—theorem holds.

Theorem 2.10 *If R is a complete relation over a finite domain S, then $R \cup R^2$ has a greatest element.*

Proof A greatest element of $R \cup R^2$ is an element x such that for any y either x dominates y or there is an s such that x dominates s and s dominates y. Again, we consider the count $C(x)$ of elements of S dominated by x and then an x with maximal count. Let C_x be the subset of S of elements dominated by x and D_x the subset of elements dominating x. Then, if there is an x' not dominated by x per $R \cup R^2$ then it must belong to D_x and it must not be dominated (by R) by any element of C_x. From completeness follows that then x' dominates (by R) all elements of C_x and therefore $C(x') > C(x)$. This is not possible, since $C(x)$ was assumed to be maximal. Therefore, x is a greatest element of $R \cup R^2$.

The restriction to finite domains is relevant. There is no issue in exhausting an infinite set, and the proof would not work.

For the following discussions, it seems a good idea to have a running example. We consider a set of five 5-dimensional vectors as

$$\begin{aligned} x_1 &= (9, 6, 3, 9, 1) \\ x_2 &= (4, 3, 8, 4, 9) \\ x_3 &= (2, 8, 5, 5, 3) \\ x_4 &= (5, 5, 4, 6, 1) \\ x_5 &= (6, 7, 2, 8, 3) \end{aligned}$$

There is no particular "story" behind this choice. If we want to evaluate these five vectors (seen as domain A) by Pareto dominance, we are out of luck. None of these 5 vectors dominates any other; for any pair of vectors, we easily find a larger and a smaller component. So there is no greatest element, and all five are maximal. Now someone comes up with a tempting idea: if we have no case where all components are better, why not compare by the number of larger components? For two vectors x and y we count the number l_x of components of x that are larger than the corresponding components of y, and we count the number l_y of components of y that are larger than the corresponding components of x. Whenever $l_x > l_y$ we consider x the better vector.

This relation has been considered several times in the past without giving it a special name. For convenience, we will call it Ticker dominance.

Definition 2.19 (*Ticker Dominance*) Given a domain $A \subseteq R^n$. A pair (x, y) of domain elements is said to be in ticker dominance relation (written $x \geq_{td} y$) whenever the number of elements of the set $L_{x>y} = \{i \mid 1 \leq i \leq n \wedge x_i > y_i\}$ is at least as large as the number of elements of the set $L_{x<y} = \{i \mid 1 \leq i \leq n \wedge x_i < y_i\}$.

We can easily get the asymmetric part by replacing $|L_{x>y}| \geq |L_{x<y}|$ by $|L_{x>y}| > |L_{x<y}|$. Why hasn't this relation been more seriously considered in the optimization research community? The simple answer is because it is highly cyclic. Table 2.1 shows all pairwise comparisons between these five vectors. For example, the entry 2-3 in the first cell of the second column means that when comparing the second with the first vector, there are 2 components of x_2 larger than components of x_1 and 3 are smaller. In fact, x_2 has larger values than x_1 for the indices 3 and 5. Figure 2.2, the graph representation of our relation, helps to see the properties of the relation directly. An arc from some node x_i to some node x_j indicates that $x_i \geq_{td} x_j$. We can observe a number of things from the graph.

Table 2.1 All pairwise comparisons in the Ticker dominance relation example

	x_1	x_2	x_3	x_4	x_5
x_1	–	3-2	2-3	3-1	3-2
x_2	2-3	–	3-2	2-3	2-3
x_3	3-2	2-3	–	3-2	2-2
x_4	1-3	3-2	2-3	–	1-4
x_5	2-3	3-2	2-2	4-1	–

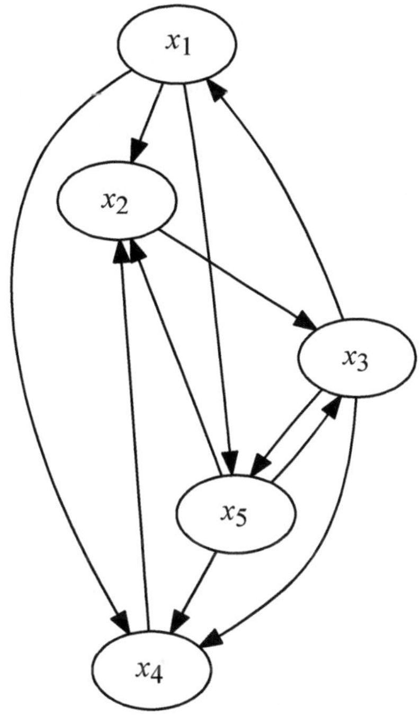

Fig. 2.2 Ticker dominance relation between the five example vectors

1. The only pair of vectors that are in mutual relation with each other are x_3 and x_5. This means that the asymmetric part is obtained by the removal of the pairs (x_3, x_5) and (x_5, x_3) from the relation.
2. No node is linked to all other nodes, so there is no greatest element.
3. Each node has at least one incident arc, even after removal of the symmetric part. Thus, the relation does not have a maximal element as well.
4. The absence of maximal elements indicates the existence of cycles. Indeed, we can find many cycles in the graph, for example $x_1 \succeq_{td} x_2 \succeq_{td} x_3 \succeq_{td} x_1$.

Now there is some hope that we can overcome the lack of greatest elements by the help of the king chicken theorem. For doing so, we need to find R^2, the relational product of our example relation by itself. It is helpful to write the relation in functional representation, as it is shown in Table 2.2. In order to compute R^2 we have to substitute

Table 2.2 Functional representation, shown are the element indices only

Index	Indices of related elements
1	1 2 4 5
2	2 3
3	1 3 4 5
4	2 4
5	2 3 4 5

each index in the right column of Table 2.2 with the indices listed in the row with that index number. For example, x_2 is in Ticker dominance relation to itself and x_3, thus the entries 2 and 3 in the second row. Now we replace 2 by the content of the second row, and 3 by the entry in the third row (which is "1 3 4 5"). So we get the indices "2 3 . 1 3 4 5" (the dot indicates the separation of the substituted groups of indices). Sorting and removing duplicates gives "1 2 3 4 5." We do the same for the other indices and receive the result

$$
\begin{aligned}
1 &: 12345 \\
2 &: 12345 \\
3 &: 12345 \\
4 &: 234 \\
5 &: 12345
\end{aligned}
$$

We do not need to unify with R itself anymore because the relation is reflexive, and R was already substituted when substituting an index with the content of the row for the same index. What immediately follows is that all but one vector is now the greatest element. With the exception of the row for x_4 all other rows contain all 5 indices. So the only thing we learned is that x_4 is not the best choice. It was already studied in [25] how the number of "king chicken" evolves under increasing randomness and problem dimension, and that there is a tendency to have more and more "king chicken"—this example is a small confirmation of this fact.

We can close the discussion on "softening" the specification of greatest elements here and consider a different approach: ranking. By this, we mean a procedure to sort the elements of a domain purely on the evaluation of the relation. In particular, extreme elements like maximal or greatest elements are assumed to be top-ranked elements among other domain elements. And we will give numbers for ranks where a lower number indicates a higher rank—this can confuse, but the idea is that there must be some "state of excellence" that cannot be beaten anymore and this suits better with a low integer number (as we use it in daily life, top quality is "A" and not "Z") while the many ways to be bad, inferior and of poor quality might be unlimited.

Now remember Theorem 2.8: if a relation R has a finite non-empty domain and is cycle-free (it means its asymmetric part is cycle-free), then the maximum set is not empty. So we can say that the maximal elements are elements of rank 1. Now we remove all rank 1 elements from the domain. If it is still not empty, the relation restricted to that smaller domain will still be cycle-free, and from the same theorem follows that there is another non-empty maximum set. A cycle within the restricted domain would be a cycle of the original domain. These maximal elements will be considered elements of rank 2. They are domain elements that are at most dominated by rank 1 elements. Now we also remove these from the domain and can find the next maximum set and rank 3 elements. And so on, until the domain is emptied by the rank-by-rank element removal. At the end we have an assignment of positive integer numbers to each domain element. Elements with a lower rank number (aka higher ranked elements) are the better elements with regard to optimality.

Just as a note: this way of ranking elements does not need transitivity of the relation, or generally a poset. It is sufficient to have a cycle-free relation. On the other hand, if we would start with the ranking of elements by making an assignment of a rank number to each domain element (without any reference to a specific relation) and then create a relation by selecting a number of pairs (x, y) with the property that in each such pair the rank of the first element is better than the second, we will have a cycle-free relation at the end—but not a transitive one. Once it was noted by Das that "people seem obsessed by transitivity" and there might be the impression that ordering of a set requires at least a transitive relation. So its kind of necessary to make the comment here that the opportunity of ranking is basically a synonym for cycle-freeness.

But as nice and simple as it seems (we will encounter a majority of cycle-free relations in the following), there is some incoherence in that way of ranking. What would happen if we were to rank the elements not "top-down" but "bottom-up"? The cycle-freeness of a relation ensures the existence of minimal elements in the same way as it ensures the existence of maximal elements. So we could start with minimal elements, give them temporary rank 1, remove them from the domain, find the minimal elements, etc. If the last rank that we find this way is k than we renumber ranks by replacing rank i with $k - i + 1$, and we also have a ranking of the domain elements, where rank 1 are the "least minimal" elements. Will this be the same ranking? Nope. Take, for example, these five vectors as the domain of a Pareto dominance relation.

$$
\begin{aligned}
x_1 &= (3, 1)\\
x_2 &= (4, 2)\\
x_3 &= (5, 3)\\
x_4 &= (1, 3)\\
x_5 &= (2, 4)
\end{aligned}
$$

We see easily that x_3 and x_5 are non-dominated and establish the set of rank 1 elements. When ignoring them, x_2 and x_4 are maximal elements, or rank 2 elements. It remains x_1 as the only rank 3 elements. But if we order in a bottom-up fashion, we find the minimal elements (which are not in a Pareto dominance relation to any other element) x_1 and x_4. After their removal, x_2 and x_5 do not dominate any other element and in the last step x_3 remains. So the ranking now is: rank 1 for x_3, rank 2 for x_2 and x_5 and rank 3 for x_1 and x_4. Magically, x_4 and x_5 have changed their ranking position by going one level down.

Maybe "magically" now was a little bit of an exaggeration. When we consider the directed graph of the relation (it is a tree), it would actually be composed of two arms that are independent with regard to dominating elements. Now, elements of the left arm are in relation to elements of the right arm and vice versa. It means essentially that we have no means to directly compare elements from the left and the right arm—then how do we come to assume that we could assign comparable rank numbers to both arms? Like the element x_4 which is "one below" x_5: why should it be of the same rank as x_2 which is "one below" x_3 and not one rank smaller?

The best way here seems to not assign rank numbers but *rank intervals*. Even if there is ambiguity in rank levels for different arms of the relation graph (as a tree graph) there are limits. Like that the rank number from above (via maximum sets) will never become larger than the rank from below (via minimum sets). So if the rank of an element is $[r_{td}, r_{bu}]$ it means that its top-down rank is r_{td} and its bottom-up rank is r_{bu}. One element is only "safely" ranked above another element when the lower end of its rank interval is still larger than the upper element of the other rank interval.

But we will soon see that this problem is part of a more grave problem with ranking in general. For now, we state that there is at least the discomfort to need a cycle-free relation. For this reason, we cannot use it to rank the ticker dominance example used before to illustrate the application of the "king chicken" theorem.

It is also possible to rank domain elements where the relation has cycles—in other words, for any relation. A famous method here is net-flow [7]. Initially, it was introduced as a special case of a better-known and more general ranking method PROMETHEE, but the simplicity of the concept behind net-flow might raise no further need to study the evolution of this concept in more detail.

What is done in net-flow, given a finite relation domain in graph representation, is to calculate for each domain element the number of outgoing arcs minus the number of incident edges. So we specify the *dominance count*

$$n_R^+(x) = |\{y \mid (x, y) \in R\}| \tag{2.15}$$

as the number of domain elements to which x is in relation R (the number of outgoing edges in graph representation) and the *dominating count*

$$n_R^-(x) = |\{y \mid (y, x) \in R\}| \tag{2.16}$$

as the number of elements that are in relation to x (the number of incident edges in the graph representation). At the moment, we postpone issues of using the asymmetric part or not, and prefer to denote the used relation attached to the symbol. In fact, both values can already be used for ranking. The larger dominance count and the smaller dominating count indicate higher ranked elements. The net-flow value appears just as a difference of both as

$$n_R^{nf}(x) = n_R^+(x) - n_R^-(x) \tag{2.17}$$

Fig. 2.3 Example where a non-greatest element has the largest net-flow value (self-relation of all elements is not shown)

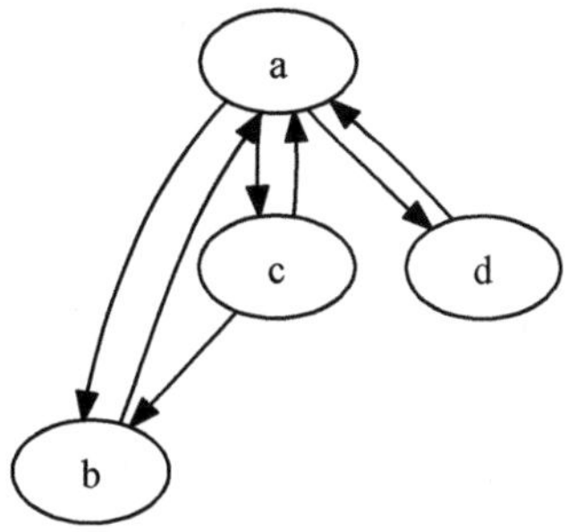

Let us see what we get for extreme elements (taking N here for the number of domain elements):

- Greatest elements are in relation to all domain elements, so the dominance count takes its maximum value N. However, there is no information on how many elements dominate a greatest element. Consider the relation shown in Fig. 2.3. The domain element a is the greatest element, and the dominance count is 4 (including itself), but there are 3 other domain elements in relation to it, so the dominating count is 3. Element a has a net-flow value of $4 - 3 = 1$. Compare this with element c: its dominance count is 3 (including itself, b, and a) and the dominating count is 1, thus the net-flow value is $3 - 1 = 2$. We also find that b has a net-flow value of $2 - 2 = 0$ and d of $2 - 1 = 1$. So the non-greatest element c has the largest net-flow value.
- Maximal elements do not have any specific values for these ranking measures. The only thing we can say is that the dominating count for $P(R)$ (instead of R) is 0. Consider now, for example, the relation $\{(a, b), (b, c), (b, d), (b, e)\}$ over the domain $\{a, b, c, d, e\}$. The maximal element a has the dominance count 1 (it is only in relation to b) while the non-maximal element b has the larger dominance count 3. Also, the net-flow of a ($1 - 0 = 1$) is smaller than the net-flow of b ($3 - 1 = 2$). Maximal elements can even have dominance counts of 0—the so-called isolated elements that do not appear in any pair of the relation.
- If the relation $P(R)$ is transitive, then the largest dominance count will be for a member of the maximum set. Each non-maximal element y must be—mediated by transitivity—dominated by at least one maximal element. So that maximal element x will dominate the same elements as y via transitivity, and it gets one more notch by dominating y. It means the maximal net-flow value will be taken by a maximal element. This is useful when the relation does not have the greatest elements.

Consider again the ticker dominance example with the relation as shown in Fig. 2.2 (where we skipped the relations of elements to themselves—there is no influence on the further evaluation). In fact, for the evaluation we only need the relation as a "connectivity pattern" and no further reference to the nature of the domain elements (like vectors in the present case) is necessary. Table 2.3 shows the results. By dominance

Table 2.3 Ranking values for the ticker dominance example.

Element	Dominance count	Dominating count	Net-flow
x_1	3	1	2
x_2	1	3	−2
x_3	3	2	1
x_4	1	3	−2
x_5	3	2	1

count, we have three winning elements: x_1, x_3, and x_5. By dominating count as well as net-flow value, there is only one winner: x_1. In all cases, x_4 (the only "non-king chicken" before) is always the loser with a smallest dominance count of 1, a largest dominating count of 3, and a net-flow value of -2 (together with x_2). A bad point is that the values are not very distinct in general, like there are two elements with the second-largest net-flow value. But as a result, there is a good indication to consider x_1 as the best choice with regard to ticker dominance. This is a nice result for an evaluation where there are neither greatest nor maximal elements and an abundance of cycles.

Actually, it appears as a disadvantage that the greatest elements do not necessarily have the largest net-flow value. However, a simple and slight modification can repair this issue. We define a net-flow* ranking measure as follows.

$$n_R^{nf*}(x) = n_R^+(x) - n_{P(R)}^-(x) \tag{2.18}$$

The change is to use the asymmetric part for the subtrahend. Since the greatest elements are also maximal elements, the subtrahend here will become 0, i.e., minimal, and the difference takes the maximum possible value for all greatest elements. In the above example, it will change the values for x_3 and x_5 from 1 to 2, so these two domain elements have the largest net-flow* value along with element x_1. For the example shown in Fig. 2.3, the asymmetric part is $\{(c, b)\}$ and thus a, c and d are maximal with dominating count 0, b with dominating count 1. The net-flow* values are now: a of 4, b of 1, c of 2 and d of 1. The greatest element a now has the largest rank according to that modified measure.

The next question is about a ranking measure where maximal elements play a bigger role. In fact, there is such a measure. Originally focusing on Pareto dominance alone, it is an integral part of the meta-heuristic algorithm SPEA2 [41] and has probably never been studied independently. But we can consider it as a suitable ranking method for any relation here. The approach follows two steps. In the first step, we calculate the dominance counts for all elements in the same manner as before. However, there is a second step to assign to each element the sum of the dominance counts of all elements that are in relation to it. This way we get the so-called S-values ("S" might indicate "strength") as

Table 2.4 S-values for the ticker dominance example

Element	Dominance count	Dominating element indices	S-value
x_1	3	3	$3 = 3$
x_2	1	1, 4, 5	$3 + 1 + 3 = 7$
x_3	3	2, 5	$1 + 3 = 4$
x_4	1	1, 3, 5	$3 + 3 + 3 = 9$
x_5	3	1, 3	$3 + 3 = 6$

$$S_R(x) = \sum_{y:(y,x)\in R} n_R^+(y) \tag{2.19}$$

A smaller S-value is better. Table 2.4 shows the evaluation for the ticker dominance example (Fig. 2.2). Again, x_1 is the winner with the smallest value of 3, and x_4 is the loser with an S-value of 9. The good thing now is that all elements have different S-values, and there is the general ranking (by index) $1 \to 3 \to 5 \to 2 \to 4$ without any ties (but note that ties are generally possible). If we take the S-values for the asymmetric part only, the picture would be a little bit different. The pair (x_3, x_5) does not belong to the asymmetric part, and therefore also x_3 and x_5 would have S-values of 3, like x_1. Maximal elements have the smallest possible S-value of 0. But then, a weak point is that it cannot distinguish between maximal elements. There is no special consideration for greatest elements by this measure.

The design of further ranking methods is up to the reader. For example, the S-value approach can be modified in a similar way like net-flow by considering the difference between dominance values of dominated elements and dominating elements. Much more elaborate methods are known, especially the well-known Analytic Hierarchy Process (AHP) method of Saaty [30].

We might ask about the case of continuous domains. Ranking here can only work if the domain is equipped with a measure (like length, area, volume). Thus, we can extend the count of dominating or dominated elements to the measure of the portion of the domain that is dominated by an element or that dominates an element. In specific circumstances, such evaluations can become quite complex. Consider the following two arguments:

1. In Chap. 1, we have extensively discussed a rate assignment problem. It appeared that all Pareto efficient solutions (i.e., the maximum set of Pareto dominance) followed the pattern $(x, 100 - x, 100 + x)$ with $0 \leq x \leq 100$. Pareto dominance-based net-flow here corresponds with the finding of the point with maximal dominated volume (it is a 3-dimensional problem). We do not need to subtract anything since these are non-dominated points. On the other hand, non-efficient elements always dominate a smaller portion of the feasible space than (Pareto-)efficient elements. The volume of the portion of the feasible space, i.e., the domain of the Pareto dominance relation that is dominated by a point (x_1, x_2, x_3) is simply

$x_1 x_2 x_3$ and thus the efficient solution with the highest net-flow is the one maximizing $x(100 - x)(100 + x)$. We have already seen this solution: it is the Nash bargaining solution, also known as the proportional fair solution. So, while primarily unrelated to Pareto dominance, it appears as the solution with the maximum net-flow of Pareto dominance.

2. If we ask about a related concept for ticker dominance we accidentally also have a simple answer for the largest net-flow value. It is because the ticker dominance has the greatest element. This is the rate allocation $(100, 0, 200)$. The condition $x_2 + x_3 \leq 200$ ensures that the last component is always the largest. Any other allocation x with $x_3 = 200$ must have $x_2 = 0$, and then everything is decided by the first value x_1. If it is not 100, then $(100, 0, 200)$ has one larger component than x while x is nowhere larger than $(100, 0, 200)$. So we can focus on $x_3 < 200$ for all other x. But now there is a similar argument about the first component x_1 restrained by $x_1 + x_2 \leq 100$, and either it is larger, or x_3 is larger, or both. So we have a single greatest element (by the way, it is the unfair solution that maximizes the total rate but gives no resource to user 2), and it is not dominated by any other solution. Therefore, it must have the maximal net-flow value as the volume of the whole domain. But the point that we want to strengthen here is different: how about other cases? For example, what is the net-flow value of the rate assignment $(50, 50, 50)$? The problem is that its calculation refers to non-trivial geometric constellations even in this simple case, and it needs a number of different cases to consider. We'd better skip the calculation for such cases like $(50, 50, 50)$ here.

So the extension to continuous domains seems possible, but can be accompanied by some notable effort and inaccessibility of general statements.

As already mentioned, even for finite domains, there is another, more general problem of ranking methods. We have already given the name: independence of irrelevant alternatives, IIA for short. Its specific meaning can vary a little, but here is the general idea: think that we are measuring temperatures at the same place at noon for a number of days. We find that the March 15 temperature was lower than the June 15 temperature. Now, no matter how long we continue the temperature value collection, this will always stay like this: the temperature on March 15 is lower than the temperature on June 15. It will also not change if we reduce our data set. Before we measured each day, and March 15 was colder than June 15, now we reduce the measurement collection to the values on the 15th of each month only. Still, the March value will be lower than the June value. Also, it does not matter how many other days March 15 appeared warmer or colder in comparison to June 15. It sounds trivial, but in fact it is a vital property of ranking: introducing or removing alternatives that are not in direct connection shall not affect the outcome.

So once we start to rank the elements of a relation domain by one of the methods given above, will this principle be valid? It will not. Here is a moderate example: The domain is $A = \{a, b, c, d, e\}$ and the relation is $R = \{(a, b), (a, c), (b, d), (e, b), (c, e)\}$. Element b has a net-flow value of -1. It is related to d, and a and b are related to it, thus $n_R^{nf} = 1 - 2 = -1$. Then, element c has net-flow value 0: it is related to e and a is related to it, $1 - 1 = 0$. So, following $0 > -1$ c is higher ranked than b.

Now we remove the element e from the domain. Compared to the temperature measurement example before, it means like deleting the measurement for some day from the data set, where the day is different from March 15 and June 15. The relation changes to $R' = \{(a, b), (a, c), (b, d)\}$. Here b has a net-flow value of $1 - 1 = 0$ and c a value $0 - 1 = -1$. Now suddenly b has a higher rank than c!

With some patience, we can find cases for the other ranking methods as well. It seems that we would have a hard time finding a ranking method suitable for any relation, and not being affected by this phenomenon.

IIA appears as a requirement in various contexts. We have already seen an interpretation of the bargaining problem solution. For a "valid" bargaining solution method, if we restrict the bargaining set such that the selected solution is still included, it should be selected for that restricted problem as well. It appears to be such a fundamental idea that, for example, in [40] it was just called *consistency*. How can we formalize IIT now for a ranking method, which is formally represented differently? In fact, there is no unique way. We will present one way now, while keeping in mind that slight modifications of the formal approach can result in completely different general statements.

1. At first, we have to qualify the "ranking method." The main point is that it is a method to assign ranking values to domain elements purely based on the relation R as a set of pairs. The standard size relation between these rank values decides whether it is a higher, same, or lower rank. The essential point is to consider it as *impartial*. It means that the ranking does not depend on some "hidden" internals of the domain elements—their nature does not affect the ranking in any way. The ranking will only depend on the specific way in which a relation links pairs of domain elements, or not.
2. There is a trivial ranking by assigning the same number to all domain elements. We will only accept it for the empty and the full relation. In fact, in these cases, all domain elements appear unqualified by the relation, or in exactly the same situation with regard to being in relation to other domain elements. Impartiality tells us that we have to assign the same rank then.
3. IIA needs reference to differing domains in order to represent the "irrelevant alternative" domain elements. But a relation over a domain is a fixed notation. When we add domain elements, we do not know how this changes the relation, as it can be extended in multiple ways. But we can make reference to subsets of the domain. If we restrict the relation to a subset of its domain, it should not affect the ranking. Actually, we can tolerate changes in the ranking number values, but not in the outcome of the comparison.
4. It is tempting to exclude ties, but we saw enough cases where ranking methods put different elements on the same rank. So we will accept these as well. When we say "it does not change the ranking" then for ties it should be interpreted as that in the restricted domain, elements of the same rank should continue to be of the same rank.
5. Last but not least: we consider finite domains here.

Given these considerations, we claim that for any ranking method that is impartial, avoids the trivial ranking except for the empty and full relations, and preserves size relations of ranking values when restricting the domain (the IIA variant we want to use now), then there is no ranking method for all relations.

If these requirements sound ambitious, there are relations where we can fulfill these demands. It is the case for a complete and transitive relation. By Theorem 2.9 we know that this ensures the existence of greatest elements. We take these as rank 1 elements and remove them from the domain. A moment of reasoning gives that the restricted relation will still be complete and transitive, and so we have another suite of greatest elements. Take them as rank 2 elements, remove them from the domain, and so on, until the domain is emptied by the repeated procedure. This ranking method fulfills IIA. The temperature example before is actually a special case of such a complete and transitive relation, like any other scalar comparison.

We can see that under the above conditions, the relation must be transitive. Actually, everything is already decided by the cases of domains with 2 or 3 elements. Take a subset of two domain elements x and y. There are four cases for the restricted relation: (1) if it is $\{(x, y), (y, x)\}$ we have to assign the same rank to x and y since they are in mutual relation to each other and there is no viable distinction between them possible—impartiality! The same for (2) if the restricted relation is empty. Also, here x and y must be of the same rank. Cases (3) and (4) are that either only (x, y) is in the restricted relation, or (y, x). Actually, we have to assign a larger and a lower rank in both cases, where the order is not relevant. If we were to assign the same rank to these elements, they would persist in being of the same rank in the original relation. But this would apply to any way of restricting the domain to a pair of its elements. It means we would end up with the trivial rank assignment since we have to assign the same rank values in all cases. This demonstrates that IIA is quite a strong requirement.

But back to transitivity. If the relation is not transitive, there must be a triple (x, y, z) of domain elements where $(x, y) \in R$ and $(y, z) \in R$ but not $(x, z) \in R$. Now we consider the restriction of R to the domain $\{x, y, z\}$. From the comments before, x and y must be of different ranks in the same direction with regard to the size of ranks. Without loss of generality, be it decreasing: $(x, y) \in R$ implies that x has a higher rank than y. In the same way, it follows that y has a higher rank than z. Now there is no pair (x, z) within the restricted relation. Then how about the pair (z, x)? If it belongs to the restricted relation (it's a cycle then), then z must be higher ranked than x, which is impossible since x has a higher rank than y and y has a higher rank than z, thus x has a higher rank than z. If (z, x) does not belong to the restricted domain, then x and z are not related at all and should have the same rank. Also, this is not possible.

So transitivity and completeness are sufficient to ensure the existence of a ranking method, and transitivity is necessary. From the latter, we can already conclude that there is no ranking method preserving IIA for all relations. Two more simple cases illustrate that transitivity or completeness alone also do not suffice to ensure the existence of a ranking method that conforms with IIA. Take the relation $\{(a, b)\}$ over a three-element domain $\{a, b, c\}$. It is indeed a transitive relation (where we actually have no "case of transitivity"). Here b is not related to c, so they must be on the same rank, but also a is not related to c and must be on the same rank as c as well. Then the ranking assignment is the same for all domain elements, but at the same time it should be a of a different rank than b. With regard to completeness, take the "circle relation" as a counterexample, i.e., the relation $\{(a, b), (b, c), (c, a)\}$ over the domain $\{a, b, c\}$. This is a complete relation, but the transitivity of the standard number larger (or smaller) relation does not allow for a consistent assignment of ranking values.

It might appear disappointing that everything about the non-existence of a suitable ranking method is already derived from the case of domains with 2 or 3 elements. We see that IIA is a strong requirement, stronger than it appears on first glance In fact, there has been a discussion about IIA going on for decades, including the approach to skip it as a requirement, replace it with a more moderate condition, or substitute it with a completely different requirement. Real life teaches us not to overweigh IIA as well. The classical "IIA story" is about the choice of dishes from a menu (as, e.g., told in [35]). One day, a customer enters a restaurant and finds three dishes on the menu. Let us say they are lunch 1, 2, and 3. The guest decides on lunch 1. After some time, the waiter comes and excuses that lunch 3 is not available at this time. Then the guest decides to take lunch 2 instead of 1. This appears as irrational behavior and the classical manner of not following IIA (where lunch 3 has not been elected anyway). But how about this? On one day, there are 20 lunches on the menu, and the customer decides for lunch 1. The next day, there are only 19 choices for a lunch menu, as lunch 3 has been removed (so lunch 1 is still available). This time the customer decides for lunch 2. Is it irrational?

2.5 Cycles of Relations

The definition of maximal elements can be applied to any relation, no matter what is the nature of the domain elements or the specific way of pair selection. However, cycles appear as a major obstacle. If the relation (as well as the alternative representation as a directed graph) is cycle-free, at least for finite domains, the maximum sets will be non-empty. But in case there are cycles, it does not mean that they truly affect the selection of maximal elements. We will discuss related issues in this section in more detail.

At first, we have to note that it is not always easy and straightforward to know whether a relation has cycles or not. We will introduce a rather uncommon relation, and for its kind of awkward way of comparing numbers, we will call it the Twister relation. Here are the definitions.

Definition 2.20 (*Twister Relation*) Given a domain A composed of pairs (a, p_a) where a is a real number and $p_a : R \to R$ is a real-valued monotonically increasing function. The number a is called the value of the pair, and p_a the moderator function of the pair. A domain element (a, p_a) is in twister relation to (b, p_b) (written as $(a, p_a) \geq_t (b, p_b)$) if and only if $p_b(a) \geq p_a(b)$.

It is also an example of a relation where the domain is constructed from composed elements. Nothing to worry about, all definitions so far apply here as well. We can compare a pair $a = (7, 2x)$ with another pair $b = (3, x^2)$ by applying the moderator function of b to the value of a, which gives $7^2 = 49$ and comparing this with the result of applying the moderator function of a to the value of b. Here it is $2 \cdot 3 = 6$. Since $49 > 6$ it follows that $a \geq_t b$.

The relation is probably more of an academic exercise, as few practical contexts might come to mind where someone attempts to compare items in that way. A less friendly interpretation would be that of an agent's defense and aggression potential. The value represents defense capacity, while the moderator function is the aggression potential—that will be imposed to an opponent in a competition. But we will not deepen that aspect here. What is more puzzling at the moment is the question: while it is a fair way to compare any pair of domain elements, in which sense is it also maximizing and ranking domain elements?

It might appear surprising that for some relevant classes of moderating functions, this relation is indeed complete and transitive, and thus a no-brainer with regard to ranking by greatest elements. We take one class where the moderator functions are multiplications by weights, written as $(a, w_a \cdot x)$. Let us call them "weighting twister relation" and simplify the notation by writing (a, w_a) for domain elements.

Theorem 2.11 *The weighting twister relation, where all weights are positive, is a complete, transitive, and cycle-free relation.*

Proof Completeness is easy to see. Take any two domain elements (a, w_a) and (b, w_b). If we assume there is no relation between both of them, it must be $w_b a < w_a b$ (so a is not in relation to b) and $w_a b < w_b a$ (b is not in relation to a). Obviously, this is not possible.

Now consider transitivity. Assume a triple (x_1, x_2, x_3) of domain elements and related weights (w_1, w_2, w_3) where $(x_1, w_1) \geq_t (x_2, w_2)$ and $(x_2, w_2) \geq_t (x_3, w_3)$. The first relation means $w_2 x_1 \geq w_1 x_2$ and the second $w_3 x_2 \geq w_2 x_3$. Multiplying the first inequality by (positive) w_3 and substituting the second inequality gives

$$w_2 w_3 x_1 \geq w_1 w_3 x_2 \geq w_1 w_2 x_3 \tag{2.20}$$

Dividing by w_2 then gives $w_3 x_1 \geq w_1 x_3$ which means $(x_1, w_1) \geq_t (x_3, w_3)$ and the relation is transitive.

Before we can also demonstrate cycle-freeness we need to clarify the asymmetric part of this relation. What are the elements of the symmetric part? We have all pairs (x, y) of the relation where $x \geq_t y$ and $y \geq_t x$. Using the notation $x = (x', w_x)$ and $y = (y', w_y)$ it means $w_y x' \geq w_x y'$ and $w_x y' \geq w_y x'$ which implies $w_x y' = w_y x'$. So the asymmetric part is composed of such x and y where $w_y x' \geq w_x y'$ but equality $w_y x' = w_x y'$ does not hold. It follows that these are the pairs where $w_y x' > w_x y'$, i.e., we just have to replace the "$\geq$" in the definition by "$>$."

Table 2.5 Testing pairwise weighting twister relations for the domain elements (1, 4), (2, 2), (3, 1) and (4, 3)

	(1, 4)	(2, 2)	(3, 1)	(4, 3)
(1, 4)	$4 \cdot 1 = 4 \cdot 1$	$2 \cdot 1 < 4 \cdot 2$	$1 \cdot 1 < 4 \cdot 3$	$3 \cdot 1 < 4 \cdot 4$
(2, 2)	$4 \cdot 2 > 2 \cdot 1$	$2 \cdot 2 = 2 \cdot 2$	$1 \cdot 2 < 2 \cdot 3$	$3 \cdot 2 < 2 \cdot 4$
(3, 1)	$4 \cdot 3 > 1 \cdot 1$	$2 \cdot 3 > 1 \cdot 2$	$1 \cdot 3 = 1 \cdot 3$	$3 \cdot 3 > 1 \cdot 4$
(4, 3)	$4 \cdot 4 > 3 \cdot 1$	$2 \cdot 4 > 3 \cdot 2$	$1 \cdot 4 < 3 \cdot 3$	$3 \cdot 4 = 3 \cdot 4$

Now it is rather easy to see that there are no cycles (of the asymmetric part). Remember that the asymmetric part is a subset of the relation, so $x >_t y$ implies $x \geq_t y$. Assume we have a chain of k related elements $x_i = (x_i', w_i)$

$$x_1 >_t x_2 >_t >_t \ldots >_t x_k >_t x_1 \tag{2.21}$$

then by transitivity follows $x_1 \geq_t x_k$ (it is worth to note that by transitivity does *not* follow $x_1 >_t x_k$). Including $x_k >_t x_1$ means $w_k x_1 \geq w_1 x_k > w_k x_1$ which is impossible.

By Theorem 2.9 we know that for a complete and transitive relation over a finite domain, it has a greatest element. From the remarks in the foregoing section, we also know that it allows us to rank all domain elements by successively removing the greatest elements from the domain and taking the greatest elements of the remaining domain as next-rank elements. One example might illustrate this for the (weighting) twister relation. Take the four pairs (1, 4), (2, 2), (3, 1), (4, 3). Then we find relations by examining Table 2.5 as follows.

$$\begin{aligned}
&(1, 4) \geq_t (1, 4)\\
&(2, 2) \geq_t (1, 4), (2, 2) \geq_t (2, 2)\\
&(3, 1) \geq_t (1, 4), (3, 1) \geq_t (2, 2), (3, 1) \geq_t (3, 1), (3, 1) \geq_t (4, 3)\\
&(4, 3) \geq_t (1, 4), (4, 3) \geq_t (2, 2), (4, 3) \geq_t (3, 1)
\end{aligned}$$

and (3, 1) appears as a greatest element. Without (3, 1) the pair (4, 3) is a greatest element, then (2, 2) and (1, 4) is least. It follows the ranking

$$(3, 1) \rightarrow (4, 3) \rightarrow (2, 2) \rightarrow (1, 4) \tag{2.22}$$

We see that even with its unconventional definition, the twister relation can be used for sorting real values (each equipped with a moderating weight) in a rather normal manner. Another case is when using power functions as moderator functions with a positive exponent. The proof that the twister relation is also complete, transitive, and cycle-free is basically the same.

But for a "free choice" of moderator functions (at least monotonically increasing functions), is it ever possible to have a cycle? The expectation of the reader might be that there should be a cycle somewhere; we just need to find suitable functions. Such problems will appear rather often while studying specific relations, and now we have the opportunity to demonstrate a "cycle construction" technique. We consider that if cycles exist, there will also be cycles of length 3. So we define the task of finding 3 numbers x_1, x_2, x_3 along with 3 moderator functions p_1, p_2, p_3 such that

$$(x_1, p_1) >_t (x_2, p_2) >_t (x_3, p_3) >_t (x_1, p_1) \tag{2.23}$$

What we note is that in order to ensure these inequalities, the number of needed function evaluations of the moderator functions is rather small. The three inequalities translate as

$$\begin{aligned} p_2(x_1) &> p_1(x_2) \\ p_3(x_2) &> p_2(x_3) \\ p_1(x_3) &> p_3(x_1) \end{aligned}$$

Thus, in total, only six different p-function values appear in these inequalities. Now we also assume that the three numbers are sorted in increasing order $x_1 < x_2 < x_3$. Since we assume monotonically increasing p-functions, this implies three more inequalities.

$$\begin{aligned} p_1(x_3) &> p_1(x_2) \\ p_2(x_3) &> p_2(x_1) \\ p_3(x_2) &> p_3(x_1) \end{aligned}$$

In Fig. 2.4, the six function values are indicated by encircled labels, and an arrow from one circle to another one indicates that the corresponding function value of the former position has to be larger than the function value at the latter position. So the task of finding a cycle translates into the task of finding an assignment of numbers to the labels such that these six inequalities hold. We see that the arrow-connection pattern does not have any cycles (in which case such an assignment would not be possible). Figure 2.5 displays the size relations between the six function values as a

Fig. 2.4 Required size relations for constructing a cycle of the twister relation

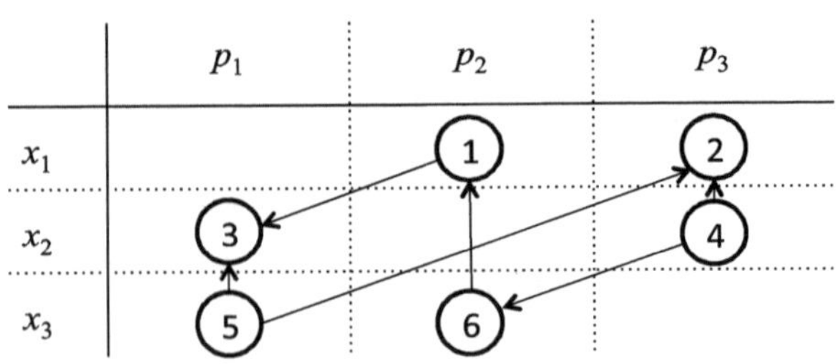

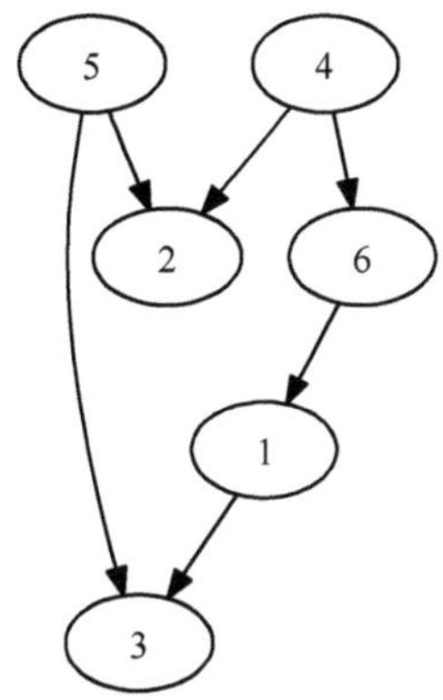

Fig. 2.5 Required size relations depicted as directed graph

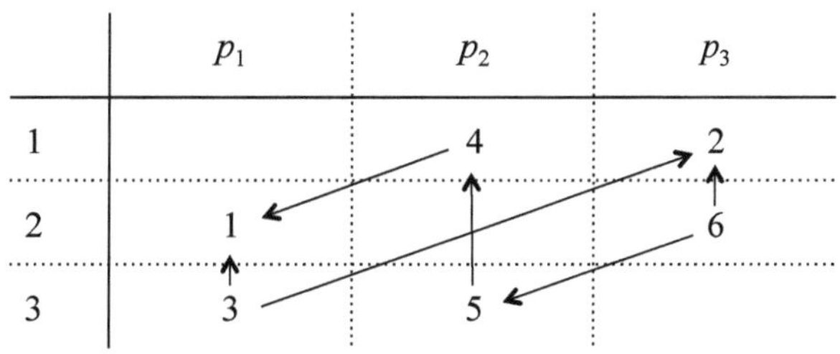

Fig. 2.6 Example numbers that fulfill the requested size relations for a twister relation cycle

directed graph. The fact that there is no cycle can be seen much more easily here. It also helps to find a "recipe" to fill in fitting numbers.

1. Select values for labels 3 and 2 freely (*for example 1 and 2*).
2. Select a value for label 5 that is larger than both values selected in the step before (*for example 3*).
3. Select a value for label 1 that is larger than the value of label 3 (*for example 4*).
4. Select a value for label 6 that is larger than the value of label 1 (*for example 5*).
5. Select a value for label 4 that is larger than the values of labels 2 and 6 (*for example 6*).

At the same time, we decide to use $x_1 = 1, x_2 = 2$ and $x_3 = 3$. Figure 2.6 shows the same scheme as Fig. 2.4 with the labels replaced by the example values. All size relations are fulfilled by this choice. Now we are ready to specify the moderator functions. In the case of p_1, we see that it is determined at two positions: for $x_2 = 2$ and $x_3 = 3$. We can use linear functions of general form $p_1(x) = ax + b$ and know that $2a + b = 1$ and $3a + b = 3$. It follows $a = 2$ and $b = -3$. Thus $p_1(x) = 2x - 3$. In a similar way, we find the other two moderator functions, giving:

$$p_1(x) = 2x - 3$$

$$p_2(x) = \frac{x + 7}{2}$$

$$p_3(x) = 4x - 2$$

It seems that we found a cycle of the twister relation, summarized as follows.

$$(1, 2x - 3) >_t (2, (x + 7)/2) >_t (3, 4x - 2) >_t (1, 2x - 3) \tag{2.24}$$

A final check: $(1 + 7)/2 = 4 > 2 \cdot 2 - 3 = 1$ confirms $(1, 2x - 3) >_t (2, (x + 7)/2)$, $4 \cdot 2 - 2 = 6 > (3 + 7)/2 = 5$ confirms $(2, (x + 7)/2) >_t (3, 4x - 2)$ and $2 \cdot 3 - 3 = 3 > 4 \cdot 1 - 2 = 2$ confirms $(3, 4x - 2) >_t (1, 2x - 3)$. Without the bias values in the linear p-functions, there would not be any cycle, but with the bias values, cycles appear.

A remark on the methodology: if there were any cycles in the scheme in Fig. 2.4 a proper value assignment would be impossible. In such a case, we would have shown the general impossibility of cycles of length 3. The assumed order of the values x_1, x_2 and x_3 does not have any influence, since we could always renumber the p-functions. But we are not finished with cycles. Excluding 3-cycles leaves us with the daunting task of showing the same for cycles of length 4, 5, etc., until we hopefully find some general argument for cycles of any length. Now thinking we are lucky that our examination could stop at cycle length 3 already: it might be considered that relations having cycles of length 4 or more, but not 3 are not so likely to appear. Yes, we can design such relations easily (just take the relation $\{(a, b), (b, c), (c, d), (d, a))\}$ over domain $\{a, b, c, d\}$), but such academic cases are rather unlikely to appear in practice—especially to appear as result of a predicate representation of a relation, i.e., the way we will use them most often.

Once we found a cycle of the twister relation, we still do not know if such cycles are abundant, average, or rare. In general, the chance to pick three elements in cyclic relation depends much on the choice of the domain. We can provide a small test. As domain we choose the set of pairs (i, p_j) where $i = 1, 2, \ldots, 10$ and $j = 1, 2, 3$. This domain has 30 elements, 3 choices for each of the ten possible values. Now we sample all possible selections of three elements from this domain ($30^3 = 27000$ possibilities) and count how often they establish a 3-cycle. The result is that there are only 72 cycles out of the 27000 selections.

Here are the 24 base cases (each cycle appears three times due to different starting points): (1, 2, 3), (1, 2, 4), (1, 3, 3), (1, 3, 4), (1, 3, 5), (1, 3, 6), (1, 3, 7), (1, 3, 8), (1, 3, 9), (1, 3, 10), (2, 3, 5), (2, 3, 6), (2, 3, 7), (2, 3, 8), (2, 3, 9), (2, 3, 10), (3, 3, 7), (3, 3, 8), (3, 3, 9), (3, 3, 10), (4, 3, 9), (4, 3, 10), (4, 4, 9), (4, 4, 10). These are only the number values; the moderator functions are p_1 for the first, p_2 for the second, and p_3 for the third elements of the triples. Among them, we find our (1, 2, 3) counterexample that we have designed before. In total, it means the chance to accidentally meet a cycle is rather low, about $72/27000 \approx 0.0027$ or 0.27%. A closer look at the list also shows that simple rules apply, for example, domain elements composed of a value larger than 4 and the moderator function p_1 do not appear in any cycle.

As an additional result, we can also count the 4-cycles. There are $30^4 = 810{,}000$ possible selections of four domain elements. Among them, 304 are 4-cycles, accounting for a mere 0.037%. We can see that the fact that a relation is not cycle-free might not be the end of the days—with regard to optimality. It still has to happen that the cycle is among the top elements before it can affect the selection of maximal elements. However, if we want to perform a full ranking of the domain by maximality, any cycle is poison. At one point during the ranking procedure, we will encounter a cycle and have no maximal elements for the next step to remove from the domain.

If cycles are not so frequent, isn't it possible then to "patch" the relation? At least in the case of a finite domain, where a representation of a relation as a directed graph is admissible, there are some related approaches. Here we will consider two possible approaches.

For the first approach, we have to revise our naming convention "cycle" a little bit. We assume a cycle as a sequence of domain elements that forms a closed loop. However, in addition to have a relation between the last and first element of the sequence, there can be relations among other elements as well. This gives further closed loops. The reader is invited to have again a look at Fig. 2.2 for noting the larger number of cycles within that graph.

From a graph-theoretic point of view, cycles belong to so-called *strongly connected components* (SCC) of a directed graph. To define things in a clean way, we denote a general graph as a pair (V, E) of vertices and edges (for a directed graph, edges are also sometimes called arcs). Not to mention that vertices are also often called nodes. There, an edge is a pair (u, v) of vertices—that way the graph represents a relation with domain V as a set of pairs $E \subseteq V \times V$. Now, a path of a directed graph is any sequence of vertices linked by edges. It means for an ordered set of k vertices v_i for $i = 1, \ldots, (k-1)$ all edges $(v_i, v_{i+1}) \in E$. Note that not all vertices in a path need to be different. In case $v_k = v_1$ we have a cycle as a special case of a path. If there is some path from a vertex v to some vertex u it is said that u is *reachable* from v. Last thing we need: a graph is *restricted* to some subset W of vertices by taking a graph with that vertex subset W as vertices and keeping only edges in E where both endpoints are in W. Then we define:

Definition 2.21 (*Connected Component*) Given a graph $G = (V, E)$. A subset $W \subseteq V$ of vertices is called a connected component of G if, within the restriction of G by W every vertex of W is reachable from every other vertex of W.

It is a *strongly connected component* (SCC) if no other vertex of G can be added to that strongly connected component W without losing the property that every vertex is reachable from every other vertex.

What if a graph is cycle-free? It does not mean that there are no SCC, but that every vertex establishes its own SCC. In general, the vertices of a graph can be split into several SCC. We can see this by using an iterative approach. We find some SCC of a graph and remove all vertices that belong to that SCC (or: if that SCC is the vertex set W, then we restrict G to $V \setminus W$). From that graph, we select another SCC, remove it, find an SCC of the remaining graph, etc. The decomposition of a graph into a set of SCCs is such that every SCC is non-empty, any pair of SCCs is disjoint, and their union gives the total set of vertices V of the original graph. Formally, for two vertices to belong to the same SCC appears as an equivalence relation.

With that, formal concepts let us go back to the original idea of "patching" a relation to make it cycle-free. What we do—while representing the relation as a graph—is to unite all elements of an SCC into a single "meta-vertex." We connect these meta-vertices by the following rule: there is an edge from SCC1 to SCC2 whenever there is at least one vertex of SCC1 linked to some vertex in SCC2. We obtain a graph that is cycle-free. It is called the *condensation* of G. So if we interpret the graph of that SCC as a binary relation between subsets of the domain (such a subset being the elements of an SCC), we obtain a relation that is cycle-free. It makes sense to consider domain elements of a cycle as equivalent: we have seen that belonging to the same SCC is an equivalence relation.

There are several algorithms to find the SCCs of a directed graph. The most famous algorithm is Kosaraju's algorithm. It has been presented by Micha Sharir in [34] based on an unpublished idea of S. Rao Kosaraju. Finding SCCs is an example for a graph-traversing algorithm strategy called depth-first search. Starting from some root vertex, the goal is to do some processing for each vertex that is reachable from that root vertex. For some vertex v, the processing goes over all vertices to which there is a direct edge from v (so-called successors of v) and finishes processing for each successor (which includes the processing of each successor of that successor) before going to the next successor vertex. This is, for example, different from the other strategy *breadth-first search* where all successors of a vertex are processed, and during the processing, encountered successors are put on a stack that serves the next processing vertex. Then the goal here is to empty the stack. In depth-first search, there is no need for a stack.

Kosaraju's algorithm is a form of depth-first search which needs two passes for finding all SCCs. A more elegant approach is Tarjan's algorithm, also based on depth-first search, and which needs only one pass. It was first published in [36] (where the algorithm is presented as an example of depth-first search, promoting the further application of this graph-traversing strategy). The idea behind this algorithm is to attempt to construct a tree from some (freely selected) start vertex to the bottom, with all vertices of the directed graph, and preserving all edges. Of course, if there are cycles in the graph at one point, one will meet an edge that links back to the tree. The algorithm keeps track of these backlinks by assigning an INDEX number to each vertex (in the order in which the vertices are sampled) and a so-called LOWLINK number. Initially, this is the same as INDEX. But whenever during the search it is found that one successor links to a vertex with a lower index, the LOWLINK value is set to that lower index. Whenever all successors of a vertex have been searched (this search includes the search within their successors, the successor of these successors, etc.) and the LOWLINK index has not changed—as a matter of fact—this vertex must be the starting point of an SCC in the graph. For keeping track of searched vertices, the algorithm maintains a stack that keeps all vertices not assigned to an SCC so far. Once a vertex is found where INDEX is still equal to LOWINDEX after searching all its successors (which means that the vertex has no link reaching into higher parts of the tree), all vertices on the stack down to that vertex establish an SCC.

We present the approach more formally. In the outer loop, the algorithm is initialized by setting the stack $S = \emptyset$ and the *index* to 0. At the beginning, all vertices also have an undefined INDEX value. Now the processing loops over all vertices with undefined INDEX—keeping in mind that the processing of some vertex will result in setting the index values of subsequent vertices, so it is not a loop over all vertices. The *processing* of a vertex v without INDEX goes as follows:

1. The INDEX of v is set to the global value of *index*, as well as its LOWLINK value. The *index* is incremented by one, and v is pushed onto the stack S.
2. Now all successors w of v are considered (i.e., (v, w) is an edge of the graph). There are two cases that can happen: the first is that w has no INDEX yet, i.e., it is met for the first time during the search. The *processing* is performed for w now, and once it returns and happens to bring with it a lower LOWLINK value than v, the LOWLINK value of v is replaced by that of w. The second case is that w is already in the stack S, so it must have been encountered during the search before the current v was processed. Then the LOWLINK of v is set to the INDEX of w. This is the event that must actually happen before a vertex gets a LOWLINK value smaller than INDEX, and which triggers the first case later on.
3. Now, when all successors of v have been processed, and it happens that still INDEX and LOWLINK of v are the same, then all vertices on the stack starting with v to the end establish a single SCC of the graph. They are popped from the stack, and the SCC is output somehow.

Since the reasoning behind this algorithm might appear rather compact, a small example might help to clarify it a bit more. Consider the directed graph shown in Fig. 2.7. The vertices are labeled by numbers from 1 to 5, and the two values in square brackets are INDEX and LOWLINK for each vertex for two processing stages of the algorithm (top and bottom parts of the figure). At the beginning, the depth-first search follows links from vertex 1 to 5. Vertices 1, 2, 3 do have only one successor, so the progress here goes directly to link 4 that has two successors. We assume that the link to vertex 5 was processed first (in fact, the order of searching successors does not influence the result of the algorithm). At vertex 5 there is no successor, so the depth-first search comes to a temporary halt. At this time, the vertex numbers on the stack are $S = \{1, 2, 3, 4, 5\}$ (right-most value was pushed last). Vertex 5, for processing no successors, does have the same INDEX and LOWLINK value of 5 and is thus the starting point of an SCC. Everything from 5 on the stack is one SCC and is removed from the stack. This is actually the single-element set $\{5\}$. Figure 2.7 indicates this by the star "*" following the square bracket.

Now the stack is $S = \{1, 2, 3, 4\}$. Vertex 4 did not finish its processing, since it also has to process the successor 2. But 2 is already in the stack! So we have to take note of the fact that vertex 4 is linked to a vertex that was processed before vertex 4 by setting its LOWLINK to the index of vertex 2—which is the value 2. This is case 2 in processing step 2 above. Then the processing returns from vertex 4 as last (and only) successor of vertex 3 to vertex 3 where there is no further edge to process. Vertex 3 notes that vertex 4 has a smaller LOWLINK than itself, so it means there is one of vertex 3 successors (or more precisely the subtree starting at vertex 3) that has a link to a vertex found before vertex 3. Thus, also vertex 3 changes its LOWLINK value to 2, copying the LOWLINK from successor vertex 4. Now, vertex 2 has finished processing its last and only successor 3. The LOWLINK of vertex 3 is 2 so there is no need to lower the LOWLINK value of vertex 2. After this, INDEX and LOWLINK

Fig. 2.7 Example for the application of Tarjan's algorithm

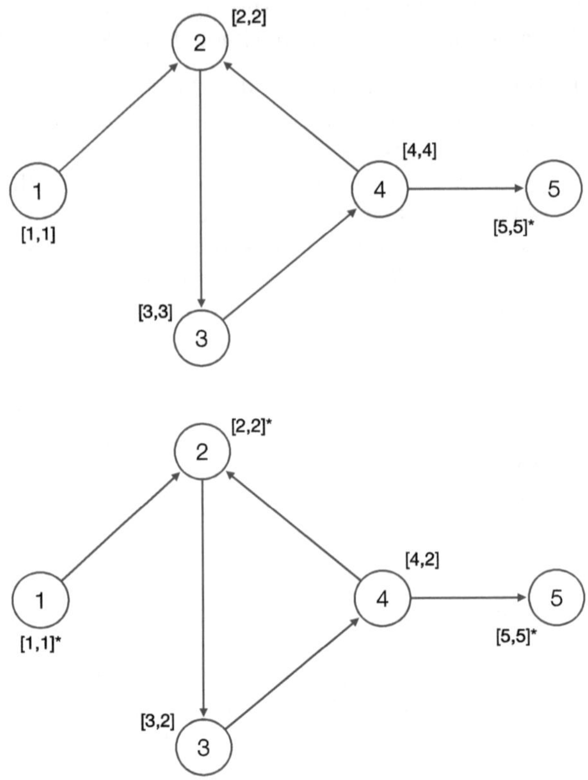

of vertex 2 are both 2. It means vertex 2 is the starting point in the stack of an SCC. We can output the next SCC {2, 3, 4} and the stack becomes $S = \{1\}$. Well, now also the processing backtracks to vertex 1 where there is no other edge to process, and vertex 2 has a larger LOWLINK than vertex 1. So also vertex 1 has equal INDEX and LOWLINK values and is the starting point of a third SCC in the graph. Since only 1 remained in the stack, this SCC is {1}. The algorithm finishes. The state at the end of the algorithm is shown in the lower part of Fig. 2.7. The graph has 3 SCC, two are single-element sets for vertex 1 and vertex 5, respectively, the third is composed of the vertices 2, 3 and 4 that obviously are a cycle in the graph.

We will stop here to go into more details about such graph algorithms. What can we gain with regard to the maximum sets of relations that have cycles? For the example shown in Fig. 2.7, we have the element with label 1 as only maximal element before and after condensation (note that the graph shows an asymmetric relation). But before condensation, we could not rank the domain elements. Now we have element 1 on rank 1, the meta-element (2, 3, 4) on rank 2 and element 5 on rank 3.

Going back to the ticker dominance example shown in Fig. 2.2, it does not gain much. In fact, all 5 vectors establish a single SCC in the graph of the relation. Remember that when considering cycles, we are only interested in the asymmetric part of the relation. In this case, both links between x_3 and x_5 have to be removed in order to get the asymmetric part.

- x_1 can directly reach x_2, x_4 and x_5. It can reach x_3 via x_2.
- x_2 can reach x_1 via x_3, directly reach x_3, reach x_4 via x_3 and x_5 via x_3 and x_1.
- x_3 can reach x_1 and x_4 directly, x_2 via x_1, x_5 via x_1.
- x_4 and x_5 can reach x_2 and from there all other.
- All elements can reach themselves via any different element.

So the single meta-element after condensation here is composed of x_1 to x_5, the whole domain of the relation, and therefore it is maximal. Instead of having no maximal element to select, we can now select either element, which is not much of an advantage. Nevertheless, we will later see a case where the condensation is useful for obtaining a result with regard to maximality.

So far we discussed one method of "patching" a relation that is potentially cyclic based on the specification of meta-elements that fuse together elements in cycles (appearing as strongly connected components of the relational graph). Another way is to remove edges or vertices from the graph—and find the order to leave the relation as "untouched" as possible, removing as few as possible.

Such ideas usually lead to combinatorial optimization problems, many of them known to be NP-hard. The removal of a minimal set of arcs from a directed graph such that the remaining graph is acyclic is known as the *Minimal Feedback Arc Set* problem. The related decision problem (for a given number k is there a subset of k or fewer arcs whose removal makes the graph acyclic?) is among the 21 classical NP-complete problems of Karp [20]. The same can be said about the task to remove as few as possible vertices from a directed graph, referring to another classical problem of NP-completeness theory: the *Minimal Feedback Vertex Set* problem. However, removing vertices appears to be a stronger manipulation than removing arcs, since its interpretation for binary relations is to remove domain elements (like declaring them infeasible) and may affect more arcs than in the related feedback arc set problem.

The reader is invited to have another look at Fig. 2.2. This graph has one connected component, so we would have to remove at least one edge from the graph in order to make it cycle-free. In fact, we can find such an edge: it is the arc linking x_2 to x_3. Remember that cycles are of interest for the symmetric part of the ticker dominance relation as shown in the graph. Thus, the mutual links between x_3 and x_5 do not count in our operation. Then, after removing (x_2, x_3), it can be seen that no cycle is left in the graph. Take the functional representation of the relation after the removal:

$$\begin{aligned} x_1 &\longrightarrow x_2, x_4, x_5 \\ x_2 &\longrightarrow \\ x_3 &\longrightarrow x_1, x_4 \\ x_4 &\longrightarrow x_2 \\ x_5 &\longrightarrow x_2, x_4 \end{aligned}$$

to also see that now x_3 is a maximal element, debunking the former favorite x_1. At the same time, we could have removed vertex x_2, which will concomitantly remove the link (x_2, x_3) as well and make the graph acyclic. But it removes all other connections to and with x_2 as well. Anyway, it is a better idea to remove x_2 than x_3, where the latter one appears as higher ranked according to various measures that we have seen already.

The disadvantage of such approaches to relation modification is that the influence on relational properties can become unpredictable. Relations may cease to be transitive or complete, and this may affect extreme values. One might come up with the idea to look for the largest transitive relation that is contained in a relation. If we do this for an asymmetric relation, or with the asymmetric part of a relation, we automatically come up with a cycle-free relation. Unfortunately, such a procedure is far from being well established. In fact, there is the concept of a *transitive interior* of a relation. It is seen as kind of opposite to the transitive closure as the largest sub-relation that is transitive. But in general, such a relation does not exist for a simple reason. The common understanding of the largest relation is that any other relation with the same property is a sub-relation of that largest relation. This cannot hold for relations that are not already transitive.

For example, take a simple cycle: the relation $R = \{(a, b), (b, c), (c, a)\}$ over the domain $A = \{a, b, c\}$. How can we remove pairs from R such that the remaining relation is transitive? Answer is that we have to remove any two of the three pairs. For example, we remove the pairs (b, c) and (c, a) and the rump relation now is $R' = \{(a, b)\}$. It might sound odd, but in fact, it is a transitive relation, for the simple reason that there are no "cases" of linked pairs. In the same way, we can reduce R to $R' = \{(b, c)\}$ or $R' = \{(c, a)\}$, all of them appear as such "transitive interiors" and there is no choice of removing a single pair for yielding a transitive relation (for example, we remove the pair (b, c) but after this the refined relation contains (c, a) and (a, b) and if it were transitive it should also contain (c, b) what is not the case). But between these three transitive interiors, none is a sub-relation of any other.

A more refined argument against the largest transitive interiors goes as follows: take any pair $(x, y) \in R$. Then the sub-relation $R' = \{(x, y)\}$ that does only contain this pair is transitive. Thus, if there were a largest transitive relation including any other transitive sub-relation, it should contain the relation R' as well. But the choice of the pair (x, y) was free, so at the end such a transitive interior would contain any pair of R. It must be equal to R.

The only way to circumvent this problem is to consider any relation R' a transitive interior that is transitive and can be obtained from R by removing the smallest number of pairs. Back to the ticker dominance example as shown in Fig. 2.2. From that graph, we want to remove as few arcs as possible in order to get the graph of a transitive relation. Again, we consider the asymmetric part only. When testing all variants, actually, we find 8 cases where the relation violates the transitivity condition. These cases are

$$(x_1, x_2), (x_2, x_3)$$
$$(x_2, x_3), (x_3, x_1)$$
$$(x_2, x_3), (x_3, x_4)$$
$$(x_3, x_1), (x_1, x_2)$$
$$(x_3, x_1), (x_1, x_5)$$
$$(x_3, x_4), (x_4, x_2)$$
$$(x_4, x_2), (x_2, x_3)$$
$$(x_5, x_2), (x_2, x_3).$$

Now, once we remove at least one pair from each of these 8 cases, the remaining relation will be transitive. In order to make the selection of pairs to remove, we follow a kind of greedy approach and observe that the pair (x_2, x_3) appears 5 times among these 8 cases. So it is a good candidate for removal. Among the remaining cases that do not contain (x_2, x_3), the pair (x_3, x_1) appears two times, and it is the next best candidate for removal. One pair remains, and we have the free choice of either removing (x_3, x_4) or (x_4, x_2). Let us take the second one, in order to avoid the removal of the domain element x_3 from any pair. We can also easily check that it is not possible to obtain a transitive relation by removing only one or two pairs. We come up with the solution to remove (x_2, x_3) (remember that in the former case, this removal alone was sufficient to make the relation cycle-free), (x_3, x_1), and (x_4, x_2). The remaining relation is shown in Fig. 2.8 and it is easily verified that this is a transitive relation. We note that in this case, the refined relation does not have a greatest element (even after adding all pairs (x_i, x_i)), while x_1 and x_3 are maximal

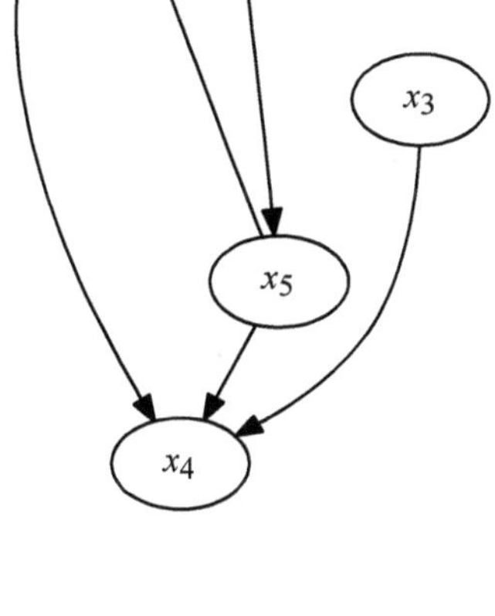

Fig. 2.8 One possible transitive interior relation for the ticker dominance example

elements. Domain element would also be maximal for the other choice of a transitive interior (removing (x_3, x_4) instead of (x_4, x_2) in the last step).

Studies on the general problematic nature of such transitive approximations are still ongoing. In [11], a mixed approach is considered where not only pairs are removed from the relation but also added in order to approach a transitive relation. In another recent contribution [27], the concept of transitive core is introduced. The author assumes a complete relation (in order to ensure some existence theorems), but the procedure to construct a transitive sub-relation is general enough. Take a relation R then the transitive core is another relation $R' \subseteq R$ with the following property: a pair (x, y) belongs to R' if for any other domain element z from $(z, x) \in R$ follows $(z, y) \in R$ and from $(y, z) \in R$ follows $(x, z) \in R$. It can be seen that the relation R' is transitive. But it is quite a strong requirement. In our ticker dominance example, more specifically its asymmetric part (but note that we do not have a complete relation here) only the pairs (x_1, x_4) and (x_5, x_4) fulfill this condition (in addition to the diagonal relation since the above condition is trivially fulfilled for $x = y$). For example, for the pair (x_1, x_4), what could be z? For one part, the choices are among all domain elements dominating x_1: this is only x_3. x_3 is also dominating x_4. For the other part, we have to look for all domain elements to which x_4 is in relation. This is only x_2. Then, also x_1 is dominating x_2. On the other hand, the pair (x_3, x_5) does not belong to the transitive core: for example, x_2 dominates x_3 but not x_5. Thus, the transitive core in this example is the set of all pairs (x, x) (the diagonal relation over that domain) and two pairs (x_1, x_4) and (x_5, x_4). We see that it has a maximum set now: all domain elements except x_4. It is the same result as for the king chicken theorem.

All the approaches to patch a relation in order to make it cycle-free share a great disadvantage: they cannot be easily applied to continuous domains (corresponding to infinite directed graphs). We want to mention one method of relation patching that can be applied independently of the nature of the domain: it is to refine the relation (with or without cycles) by logically AND-ing it with a cycle-free relation. Given a relation in predicate notation, where $(x, y) \in R$ whenever predicate $P_R(x, y)$ is true, we take another relation Q in predicate notation using predicate $P_Q(x, y)$ and which is known to be cycle-free, and specify a new relation S with the predicate $P_S(x, y) = P_R(x, y) \wedge P_Q(x, y)$ for checking if (x, y) belongs to that new relation S. We have to be careful about relational properties. Here are some facts:

- If Q is asymmetric, then S is asymmetric as well. Some pair (x, y) belongs to the symmetric part if and only if $(x, y) \in S$ and $(y, x) \in S$. It means that the following predicate is true: $P_R(x, y) \wedge P_R(y, x) \wedge P_Q(x, y) \wedge P_Q(y, x)$, which means the same as that (x, y) belongs to the symmetric part of R and to the symmetric part of Q. Or: $I(S) = I(R) \cap I(Q)$. Since Q is asymmetric, $I(Q) = \emptyset$ and therefore $I(S) = \emptyset$. The composed relation S is asymmetric. Therefore (since our final goal is cycle-freeness of the asymmetric part), it is always a good idea to combine with an asymmetric relation.
- If Q is antisymmetric then S is antisymmetric as well. We recall that antisymmetry of a relation R over a domain A means that from $(x, y) \in R$ and $(y, x) \in R$ it

follows that $x = y$. Or: $I(R) \subseteq D_A$. Using the same expression as in the item before from $(x, y) \in S$ and $(y, x) \in S$, it follows that $(x, y) \in Q$ and $(y, x) \in Q$ and therefore $x = y$ since Q is antisymmetric.

- If R and Q both are complete relations, S is not necessarily complete. If for some pair (x, y) the predicate $P_R(x, y)$ is true and $P_R(y, x)$ not, and $P_Q(x, y)$ is false and $(P_Q(y, x))$ true then we have $(x, y) \in R$, $(y, x) \in Q$ but its neither $(x, y) \in S$ nor $(y, x) \in S$.
- If R and Q both are transitive, then S is transitive as well. If we have any triple (x, y, z) where $(x, y) \in S$ and $(y, z) \in S$ it means $(x, y) \in R$ and $(y, z) \in R$ and where R is transitive it follows $(x, z) \in R$ and in the same manner we can find $(x, z) \in Q$ to follow from $(x, y) \in Q$ and $(y, z) \in Q$. So $(x, z) \in R$ and $(x, z) \in Q$ and it also follows $(x, z) \in S$.
- If Q is cycle-free (which we understand as: there are no cycles in $P(Q)$ while there might be cycles in Q itself), then S is not necessarily cycle-free. It is cycle-free in case Q is asymmetric. As a simple example that cycle-freeness of Q itself is not sufficient, take Q to be the full relation U_A over the domain $A = \{a, b, c\}$. It is cycle-free since its asymmetric part is the empty relation. However, if we have a cycle in R like $a >_R b >_R c >_R a$ then we have the same cycle in $S = R \cap Q = R$. On the other hand, if Q is asymmetric, any cycle of S would imply a cycle in Q.
- If Q is antisymmetric and cycle-free, then S is also cycle-free. We assume that S would have a cycle $x_1 >_S> x_2 >_S \ldots >_S x_k >_S x_1$ of length k. This means that the pairs $(x_1, x_2), (x_2, x_3), \ldots, (x_k, x_1)$ all belong to $P(S)$. Since $P(S) \subseteq S$, they all also belong to S, and from $S = R \cap Q$ it follows that they all belong to Q. But if they all would also belong to $P(Q)$, then Q would have a cycle. Therefore, at least one of them does not belong to $P(Q)$. Assume it is the pair (x_1, x_2) (the following argument is the same for any other pair). So $(x_1, x_2) \in Q$ which does not belong to $P(Q)$ must belong to $I(Q)$ and therefore also $(x_2, x_1) \in Q$. But Q is antisymmetric, and it follows $x_1 = x_2$. Then, the pair (x_1, x_2) cannot belong to $P(S)$, contrary to our assumption that (x_1, x_2) is part of a cycle of relation S.

We had a little effort to consider the cases where we can refine a cyclic relation by another cycle-free relation in order to get a cycle-free relation. It is generally the case if we refine by an asymmetric or antisymmetric relation. Especially, it means that we can always refine a relation by using Pareto dominance, even in weak form (since this is an antisymmetric relation). However, the approach fails if the relation is already implied by Pareto dominance. Then Pareto dominance is a subset of the relation, and the result of our operation is Pareto dominance. It is cycle-free, but we did not want to achieve cycle-freeness by removing any traces of the original relation. For example, it cannot help to make the Ticker dominance cycle-free since this is already implied by Pareto dominance. It can help to make the Twister relation cycle-free by combining with the standard size relation of real numbers: it means if we use any p_i with fitting domains for $(x_i, p_i) \geq_S (x_j, p_j)$, we require $p_j(x_i) \geq p_i(x_j)$ and $x_i \geq x_j$.

This way of designing a modified (refined) version of R, we can ensure that for a finite domain, there are maximal elements. The same is not true for the greatest elements. Even if both relations were complete and transitive before, they do not need to be so after the refinement. As another disadvantage that has already been mentioned, the combined relation might have a larger maximum set than the original one.

2.6 Social Choice and Rationalization

Relations play a central role in mathematical economics to formalize the concept of social choice. A strict application of a relational framework was provided by Kotaro Suzumura [35].

Definition 2.22 (*Social Choice Function*) Given a universe of choices as set U and a system $\mathcal{S}$ of non-empty subsets $S \subseteq U$, so-called choice sets. Then a (social) *choice function* C is a mapping $C : \mathcal{S} \to (\in^{\mathcal{U}} - \emptyset)$ such that $C(S) \subseteq S$.

This notation can represent many concepts, for example, an election: the choice set X is all electable persons of a community, and the subsets S are sets of candidates for various elections. Then, a voting scheme can be applied to such candidate sets, with the effect of selecting a winner for each contained candidate set. In this case, the subset of S usually contains exactly one element, the chosen candidate, but in other circumstances, the chosen subset can also have more than one element.

Figure 2.9 illustrates the case where the choice set X has 7 elements. Here, the choice function does not have to provide means for choices from *all* possible subsets,

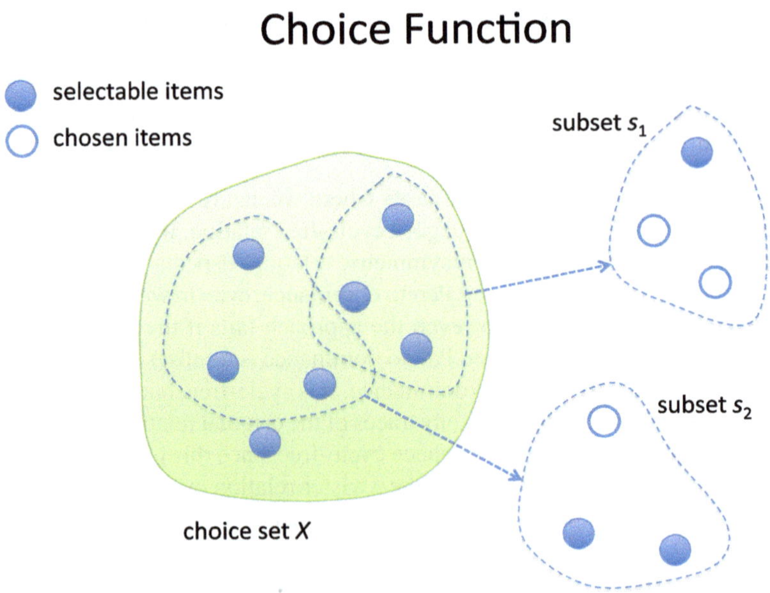

Fig. 2.9 Social choice from a choice set

but only for two subsets: it selects two items for subset s_1 and one item for subset s_2 (indicated by open circles). Note that not all elements of X need to belong to a subset, and that elements can also belong to more than one subset. This concept of social choice already appears as a rather generic abstraction. Nevertheless, a few points have to be taken into account:

- It is not required that $\mathcal{S}$ is the system of all subsets of U (excluding $\emptyset$). The question then is why an agent capable of making choices from some sets S should not be able to make choices from other sets. However, it might depend on the context of the choice. If it is an experiment with some limited set of choice sets, nothing else can be concluded about choices beyond that restricted set. But if the agent follows some procedure when evaluating a choice set, for example, a voting scheme, then it should be basically possible to apply the choice function to all non empty subsets.
- The selection from the choice set can contain more than one element. In election schemes or decision-making, this is often considered a *tie* or *trade-off*, and additional means are needed if exactly one element has to be selected. This institution is sometimes called a *decision maker*.
- The selected subsets have to be non-empty. It is sometimes questionable if agents always have to make a choice, even if the presented choice set contains only unattractive items. So this requirement is sometimes restrictive.
- It is not said who makes the choice. There can be a difference if one agents make a choice, or a group of agents. The term social choice might rather refer to a group of agents.
- Also, nothing is said about the modality of presenting the choice set. Especially, the choices of a choice set can be presented one by one, and the agent has to decide whether to keep the offered item, while evaluating against already offered or kept items, and also to possibly remove from kept items if a newly presented item appears to be more attractive. The other way is to provide all choices at once, and the agent (or the agents) acts as an oracle to come up with a subset of choices without revealing any rationale. These are the extreme cases, but many other manners lie in between, e.g., pairwise tournaments.

A rationalization of a social choice is a formal way of regenerating the choice by means of a relation. It can also be said that we are seeking an explanation or rationale behind the contingent selections made by the agent or by the group of agents. We are assuming that we can find such a rationale by a hidden ability of the agent(s) to compare items in a pairwise manner. Using relations, as we have seen already, allows for a variety of specifying subsets from its domain. The way followed in [35] is:

Definition 2.23 Given a social choice function C over the universe U and system $\mathcal{S}$ of choice sets. Then C is called rationalizable if there is a relation R with domain U such that for each $S \in \mathcal{S}$ the choice $C(S) = B_{R\downarrow S}$.

This means the choices from S coincide with the best elements of S according to a relation R over U when restricted to the choice set S. It is also called G-rationalizability of the choice function C. Figure 2.10 illustrates a *rationalization*

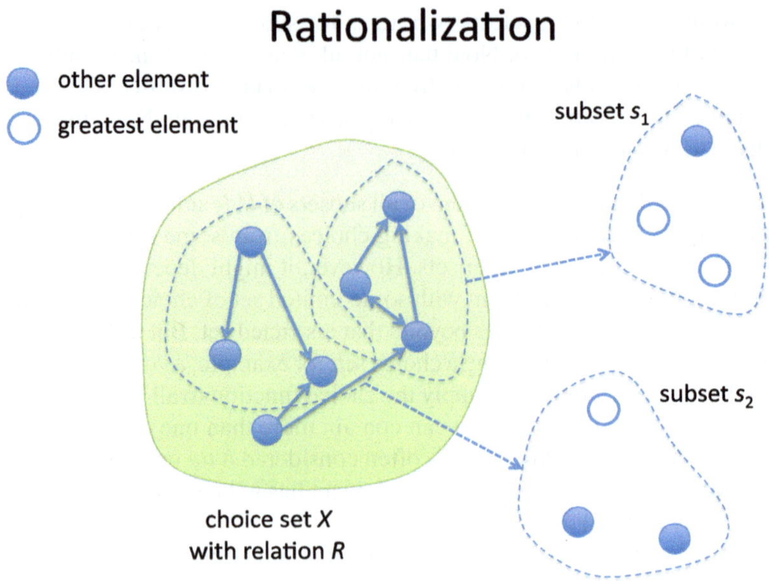

Fig. 2.10 Rationalizability of a social choice function

of the choice function that was shown in Fig. 2.9 by providing a correspondence between choices from subsets and greatest elements (it also contains a case where a greatest element is not maximal, but chosen nevertheless).

In general, not all choice functions will have a rationalizing relation, and there can be more than one rationalization of the same choice function.

Examples for this can be found in [35]. We take the item universe as $U = \{a, b, c\}$ and the system $\mathcal{S}$ contains two sets: $S_1 = \{a, b\}$ and $S_2 = U$. A choice function C selects from S_1 the choice set $C(S_1) = \{a\}$ and from S_2 the choice set $C(S_2) = S_2 = U$. Now assume there is an R rationalizing this choice. Since it selects a from S_1 but not b we know that $R \downarrow_{S_1}$ contains (a, a) and (a, b) but not (b, a) or (b, b). On the other hand, since $b \in C(S_2)$ it must be $(b, b) \in R$ and $(b, a) \in R$, a contradiction. To see that a rationalizing relation is not unique, just take a rationalization and add some pair to R that is irrelevant with regard to best set selection.

The rationalization has an inverse problem, that of characterizing all choice functions for which a given relation, or one or more relations from a given family of relations, provides a rationalization.

We may sometimes find the best set of requirements too strong, especially when focusing on such a family of relations. Therefore, we provide another version of rationalizability, using the maximum set instead of the best set.

Definition 2.24 Given a social choice function C over the universe U and system $\mathcal{S}$ of choice sets. Then C is called M-rationalizable if there is a relation R with domain U such that for each $S \in \mathcal{S}$ the choice $C(S) = M_{R\downarrow S}$.

We refer again to Theorem 2.4 stating that $M_R = B_{R^a}$. It means if we can reproduce the social choice by maximal elements of a relation, restricted to the choice set, we can also reproduce the choices by greatest elements of the dual relation. The converse does not necessarily hold, i.e., M-rationalizability implies G-rationalizability but not vice versa.

This is also taken from [35]. We use

$$\begin{aligned} U &= \{a, b, c\} \\ \mathcal{S} &= \{\mathcal{S}_\infty, \mathcal{S}_\in, \mathcal{S}_\ni\} \\ S_1 &= \{a, b\},\ \ S_2 = \{a, c\},\ \ S_3 = \{a, b, c\} \\ C(S_1) &= S_1,\ \ C(S_2) = S_2,\ \ C(S_3) = \{a\} \end{aligned}$$

This choice is G-rational with relation

$$R = D_U \cup \{(a, b), (b, a), (a, c), (c, a)\}$$

For example, for S_2 both a and c are greatest elements, while for S_3 only a remains, since $(c, b) \notin R$. Note that $P(R) = \emptyset$. Assume that C is M-rational with some relation R'. Then from $C(S_1) = S_1$ follows

$$[(a, b) \notin P(R')] \vee [(a, b) \notin R' \wedge (b, a) \notin R']$$

For example, a is in $M_{R'}(S_1)$ (here and in the following we skip the subscript for the restriction of R' to S_1) either because (a, b) or (b, a) do not belong to $P(R')$—and then one non-membership implies the other non-membership—or because b is not in relation to a. For b, it follows in a similar way. So we can see that $(a, b) \notin P(R)$ because otherwise $(a, b) \notin R'$ would not be possible.

We conclude in a similar way from $C(S_2) = S_2$ that

$$[(a, c) \notin P(R')] \vee [(a, c) \notin R' \wedge (c, a) \notin R']$$

and see that $(a, c) \notin P(R')$. The corresponding conclusion for S_3 and $C(S_3) = \{a\}$ is a little bit longer.

$$[(a, b) \in P(R') \vee (c, b) \in P(R')] \wedge [(a, c) \in P(R') \vee (b, c) \in P(R')]$$

and states basically the reason for excluding b and c. Now since $(a, b) \notin P(R')$ and $(a, c) \notin P(R')$ this can only be true if $(c, b) \in P(R')$ and $(b, c) \in P(R')$ which is in serious conflict with the definition of the asymmetric part. We conclude that such a relation R' cannot exist.

However, many other ways can be designed to employ relations for regenerating or better "modeling" choice functions. To list a few:

- We require that $C(S)$ does not contain least or minimal elements of the rationalizing relation R restricted to S. This way, we attempt to explain the non-selection of elements.
- We require that only inner elements of the rationalizing relation $R\downarrow_S$ occur in $C(S)$. A moderate agent avoiding extremes might behave like this.
- We do not require exact correspondence but sub- or supersethood of best/maximum sets and $C(S)$.
- We consider two relations R_1, R_2 such that the choice $C(S)$ is equal to the upper approximation by R_1 of the best set of S according to R_2. This represents some additional tolerance in the choices. Corresponding modifications using lower approximations, maximum sets, and subsets or supersets of $C(S)$ can be considered as well.

Many relations between these specifications can be envisaged, but this will be beyond the scope of the current discussion. We only want to point out that the selection of G-rationalizability is one of many ways to employ relations in a model of social choice.

What is the advantage of knowing that a choice function is rationalizable? In [35] it is also demonstrated how various economic theories can be brought under the umbrella of this formal framework, by providing additional assumptions about the choice functions as so-called *axioms*. For example, Arrow's Axiom is given as

$$\forall s_1, s_2 \in S : s_1 \subset s_2 \rightarrow [s_1 \cap C(s_2) = \emptyset \vee s_1 \cap C(s_2) = C(s_1)] \tag{2.25}$$

The "meaning" of this may take a moment to comprehend. It is virtually an exclusion of choices depending on the absence of choice items. Assume a scenario where someone goes into a cafe and wants to order cake, and the waiter explains the menu choices: blueberry cake, cheese cake, and cherry cake. The customer decides on the blueberry cake. After a moment, the waiter returns to the table and excuses that he made a mistake when telling the choices, and there is no cheesecake available anymore. The customer replies "If that's the case, I will take the cherry cake!" Such a change of mind is actually excluded by Arrow's Axiom for a "rational" agent. There, s_2 is $\{cheese, cherry, blueberry\}$ and $s_1 = \{cherry, blueberry\}$. Then, the customer is the "performer" of the choice function C and this assigns $C(s_1) = \{cherry\}$ and $C(s_2) = \{blueberry\}$. So we have $s_1 \subset s_2$, that means the range of available items s_2 is narrowed down to s_1 ("no cheese cake"). But the implication appears to be wrong: $s_1 \cap C(s_2) = \{blueberry\}$ is not empty, and $s_1 \cap C(s_2) = \{blueberry\}$ is different from $C(s_1) = \{cherry\}$.

So, Arrow's Axiom reads like: if the range of available items (s_2) is narrowed down to s_1, but still contains some previously chosen items, no previously unchosen item becomes chosen and no previously chosen item becomes unchosen.

As an example for the further reasoning: based on this assumption about the choice function, and if the choice function is rationalizable by a relation R, the relation is an ordering if and only if the choice function fulfills Arrow's Axiom (see Theorem 2.2 in [35]). This is a very short and sketchy introduction to the Suzumura approach

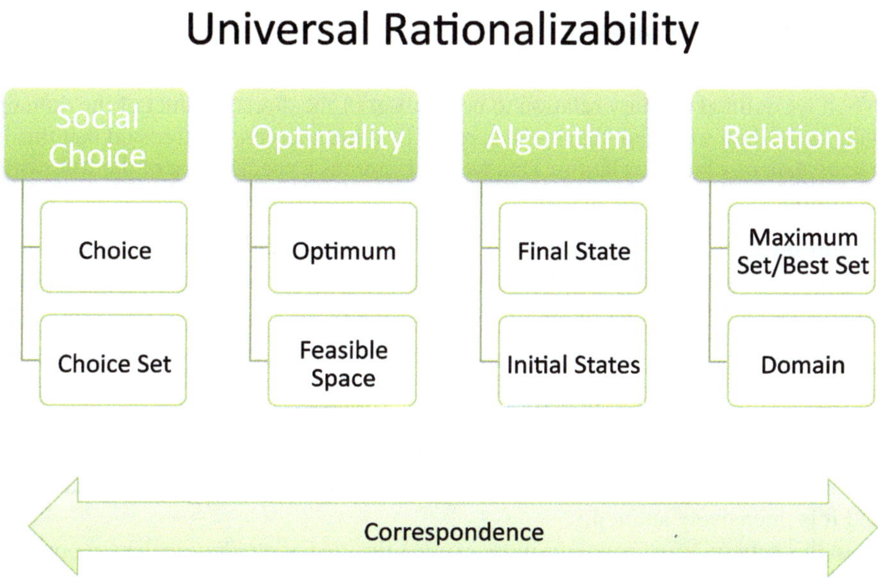

Fig. 2.11 Universal rationalizability, extending the concept of a social choice function to other domains

to social choice, and we highly recommend the book [35] for further reading. As it was stated in the introduction, our concern is optimality as a means to efficiently select configurations for the design and operation of a system. However, the pure set notation of rationalizability implies its applicability to various other contexts, and we want to call this universal rationalizability (see Fig. 2.11). Rationalizability was given by the correspondence between best or maximum sets and social choices. But also, if we want to formalize optimality, we refer to the selection of optimal states for each element of a collection of subsets of the feasible space, and we can consider its rationalization by a relation. Then, the optimal states correspond with the maximum or best sets of the relation, restricted to the corr. subset. In fact, the traditional approach to optimization is to maximize a real-valued function, so the relation here is simply the $>$-relation (alternatively $\geq$), where both, the best and maximum sets correspond with the largest element or supremum.[1] In multi-objective optimization, we refer to the Pareto-front (i.e., the maximum set of the Pareto dominance relation) of non-dominated elements in a similar sense (see next section). But the correspondence also shows that we can provide optimality by specifying any other "better-ness" relation in the same fashion.

[1] To be formally correct, we have to consider the function as providing a pre-ordering relation between the elements of its domain, and we are looking for the maximum set of this pre-order.

2.7 General Relations

When we defined a binary relation to be a subset of the direct product of the domain with itself, then we should have—according to some sources—better said that this is a *homogeneous* binary relation. This way, it becomes a special case of the following definition.

Definition 2.25 (*Heterogeneous Relation*) Given two sets A and B. Then a heterogeneous binary relation R is a subset of $A \times B$. In case $A = B$ H becomes a homogeneous binary relation, or just a binary relation.

As the binary relation appears as a special case of general heterogeneous relations, heterogeneous relations are a special case of homogeneous relations. We take as domain $A \cup B$ and add the convention that for any pair $(x, y) \in (A \cup B \times A \cup B)$ it has to be $x \in A$ and $y \in B$. Well, it is a little bit of an artificial construction, but it works. In graph theory, such a separation of the domain is called graph matching, and it is intensively studied.

In this general form, relations more express the idea of *attribution* than of optimality. The domain A can be a set of objects, and the set B of labels, colors, attributes, etc. So a relation here is an assignment of a color, label, etc., to each object. In this sense, heterogeneous relations play a major role in knowledge representation, formal languages and grammars (where they model attributes) or relational databases. A concept like asymmetry or the like does not make much sense if related pairs stem from different domains. In the following, we will not start to refer to any relation studied here as a homogeneous binary relation. This is not supposed to be an undervaluation of the role of heterogeneous relations. There are excellent sources for the deeper study of general relations, e.g., [33].

Coming next to the issue of the term "binary"—it somehow implies to consider also ternary, or more general n-ary relations. In fact, we can define

Definition 2.26 (*N-ary Relation*) Given a domain A. A subset of $A^n = \underbrace{A \times \ldots \times A}_{n \text{ times}}$ is called an n-ary relation over domain A.

In addition to binary relations (or "2-ary relations"), are n-ary relations of interest to model optimality? The answer is rather difficult. In fact, with regard to optimality, there are only a few studies on this topic. Most of these studies consider ternary relations (i.e., $n = 3$). So a ternary relation over a domain A is a subset of triples from $A \times A \times A$. The point is that such relations are convenient for the study of *betweenness*. Even then, the number of varying definitions is large. An overview of these concepts is given in [32]. Let us see an example. In [8] a ternary relation R is called a *betweenness relation* if the following two conditions hold:

1. If $(x, y, z) \in R$ and $(y, x, z) \in R$ then $x = y$.
2. If $(x, y, z) \in R$ then also $(z, y, x) \in R$.

The convenient reading for $(x, y, z) \in R$ then is that "y is between x and z." Then, y is also between z and x, which gives the second condition. The first condition says that if y is between x and z, then x itself cannot be between y and z unless $x = y$. As for binary relations, a number of properties can be defined for ternary relations. For convenience and following [8], we use the notation $x|y|z$ for a triple belonging to a betweenness relation R.

- A betweenness relation is linear if from $y \neq z$, $x|y|z$ and $y|z|u$ follows $x|z|y$ and from $x|y|z$ and $x|u|y$ follows $x|u|z$ and $u|y|z$. This refers to consistency in extending betweenness to the inside or outside.
- A betweenness relation is complete if and only if for all domain elements x, y, and z we have $x|y|z$ or $y|x|z$ or $x|z|y$, i.e., among any three domain elements at least one has to be between the other two.
- A ternary relation R is symmetric if and only if from $(x, y, z) \in R$ follows that $(z, y, x) \in R$. It is asymmetric if and only if $(x, y, z) \in R$ implies that $(z, y, x) \notin R$ and antisymmetric if and only if from $(x, y, z) \in R$ and $(z, y, x) \in R$ follows $x = z$ [28, 29].
- A ternary relation R is cyclic if and only if from $(x, y, z) \in R$ follows $(y, z, x) \in R$ (and from this in turn $(z, x, y) \in R$). It is transitive if and only if from $(x, y, z) \in R$ and $(z, y, u) \in R$ follows $(x, y, u) \in R$ [29].

All these definitions refer more or less directly to the understanding of a ternary relation as a kind of betweenness. In the specific modeling of such relations, general metrics can be used as argued in [32]. Based on this, for example, the shortest path metric of a graph can be used to define a betweenness relation. It means a vertex is between two other vertices if it lies on a shortest path connecting these vertices. One example result then is that if a betweenness relation is linear and complete it corresponds with such a graph betweenness where the graph is a tree [8].

We may wonder how this exposition relates to optimality. Of course, it does in a limited sense: assume a domain element that is between any pair of other domain elements. We may qualify this as core, center, innermost element, etc., and thus refer to some kind of extremality. But so far, this direction has not been investigated.

On the other hand, there is a notable increase in complexity, even for ternary relations, with regard to the way a relation models some aspect of reality. Remember, for binary relations, we only had to consider an aspect of similarity (the symmetric part) versus an aspect of betterness (the asymmetric part). What kind of classes would we have for a ternary relation? The answer is, following the same approach as for binary relations: we have 32 classes! If we assume an element $(x, y, z) \in R$, there are actually 32 possibilities for the permutations of (x, y, z), also belonging or not belonging to R. There are five such permutations (excluding the identity permutation)

(x, z, y)
(y, x, z)
(y, z, x)
(z, x, y)
(z, y, x)

and for each class, we had to specify exactly which of them appear together with (x, y, z) in the relation. This gives $2^5 = 32$ classes. One class is for example all

$(x, y, z) \in R$ where $(z, y, x) \in R$ as well, but not (y, x, z), (y, z, x), (z, x, y) and (x, z, y). One may note that this example fits with a betweenness relation as discussed above. But it seems hard to think about a nice and intuitive interpretation for all 32 cases, especially with regard to optimality. And in general, an n-ary relation will have $2^{n!-1}$ such classes. For $n = 2$ this gives 2, but grows rapidly with n. This might help to explain a lower concern of n-ary relations for optimality, and the general preference for binary relations.

In this regard, a small comment about the case $n = 1$. The formal meaning of a 1-ary relation over a domain A is a subset of A. It can be achieved by a predicate. Whenever the predicate is true for some domain element x, it belongs to the relation. This is actually the base for the Boolean algebra, which is a common interpretation of 1-ary relations. And it has relevance for optimality. Here, we are not considering specific elements of the domain and how they are related to all other domain elements. Instead, we specify a number of reasonable conditions for optimality. Whenever they hold, we consider the solution as optimal. We have already seen an example for this in the Nash bargaining solution in Chap. 1. The art is to define conditions such that only one or very few solutions can fulfill them all. But at the end, this may result in a binary relation like the maximal product of allocations in case of the Nash bargaining solutions, and at least both approaches appear equivalent. It cannot be said if this is generally the case.

As a last extension of relations, we will have a closer look at the edges of a graph representation of a relation. While we gave all kinds of freedom to the vertices (or domain) of a relation, an edge is just an edge. It's there to connect two domain elements, or it's not there. As we have the concept of a *weighted graph*, where some value is attached to each edge, we have the concept of a *valued relation*. This is actually a mapping $A \times A \to D$ of the set of pairs of domain elements into some value range D. A natural choice is $D = [0, 1]$ where each pair is assigned some degree, weight, or else that describes the strength to which that pair belongs to the relation. But in fact, the range can be of any nature as well. We give a few examples.

- A **time stamp** is assigned to each pair. It means that the fact that some pair (x, y) is indeed in relation is not known immediately. It needs some time to "spread the news" about the relationship within the relational graph. In this manner, relations can be used to model, for example, mobility of vertices (it might be better to call the vertices *nodes*). If two nodes come close, they can exchange information, esp. about their own relativeness, or share information about the relativeness of other, remote pairs. A similar approach is to assign **effort values**. Only when reaching that effort value as a threshold, the relativeness of some pair becomes known. It is relevant for studies where the resources to gather knowledge about a relation is limited and this limitation is an important factor of the processing.
- A **relevance weight** is assigned to each pair. This relevance weight then describes the relevance of that particular relation with respect to other pairs and the way they are related. For example, it can represent a confidence level about a particular relation. In the experimental method of pairwise comparison—esp. in psychology where it was first applied by Thurstone [37]—the goal is to unreveal a preference

relation among items of some set. The items are presented in pairs to subjects, and the preference of one or other item over the other (incl. indifference) is recorded. A pairwise matching is a table with an entry of the total number of preference choices for each pair of items. These values can serve as such weights. The famous Analytic Hierarchy Process (AHP) [31] is a method to rank the domain elements from such values.

- A **probability** is assigned to each pair. This represents the chance or frequency by which that relation occurs. It puts a relation into the general probabilistic framework and allows for corresponding evaluations.
- A **class, label or color** is assigned to each pair. It includes the case of multiple such labels. If there is exactly one label for each pair, this imposes an equivalence relation among the pairs of the relation. In this sense, we can consider the "blue part" or "red part" of a relation. It can be useful for the comparison of classes by evaluation of the interplay between the different relations.
- A **predicate** is assigned to each pair. Then, a pair (x, y) is in relation if and only if that predicate is true. It is worth noting that here, the predicate should not depend on x and y but on some external factor that justifies the relation. For example, a food preference can differ depending on the time of day. Then this is a proper way to model such a situation by a predicate-valued relation.
- There can be many other, more subtle choices for such valuations. In decision-making, a classical example is when the preference depends on the amount or quantity of items. While there is indifference between a single apple and a single orange, six apples are preferred to four oranges, while five oranges are preferred to six apples. Such information can be kept in tables or functions assigned to each pair. At the end, any kind of information storage that assists further conclusions from relativeness can be imagined.

We may ask how such a valuation of relations affect the optimality, as represented by a relation. The answer is that it much depends on the specific interpretation of the values. With reference to the above list, we can consider a few cases.

If the pairs of a relation have a time stamp, at each time instant, we can threshold the pairs by keeping all pairs having a time stamp below that instant. It means that at this moment, the pairs known to be in relation establish an (unvalued) binary relation. The extreme elements of the domain can be derived from this relation. What we have at the end are the greatest or maximal elements that evolve in time, and a dynamic picture of the optimality structure. The time stamp model can be enriched by a forgetting or evaporation factor that keeps the relativeness only in mind for some limited time. Also here the presentation of optimality is straightforward. We can say the same about valuations like symbols, classes, labels: there is a direct way to specify a binary relation by selecting some specific label and seeking its extreme values. In this manner, we can yield a "blue maximum" or "red maximum."

It is different if we assign numerical values. Then, a transition to a "crisp" binary relation is not straightforward, and the evaluation depends strongly on the way we interpret these values. Let us consider an example, as shown in Fig. 2.12. It is a valued relation over the domain $A = \{a, b, c\}$. For now, we want to ignore self-links, i.e.,

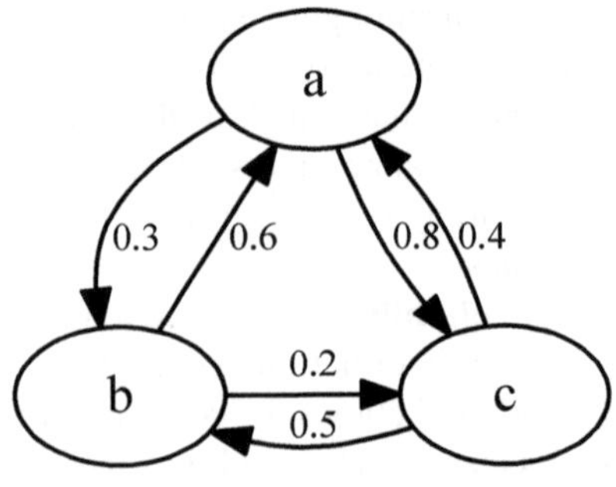

Fig. 2.12 Example for a valued relation. For simplification, the self-links are all assumed to have a value of 1

pairs like (a, a). If it matters, one can assume that their values are 1, then they will not affect the following evaluations. What we see in this graph is that the base graph is essentially a fully connected directed graph. So for each pair of domain elements, there is an arc from the first to the second, and an arc from the second to the first. Here we consider values from the range [0, 1]. It appears natural to assume that a value of 1 means that the pair fully belongs to the relation and a value of 0 with the opposite case. If we threshold all weight values, we will always get a binary relation with corresponding extrema. But this would produce a lot of extrema and be of little merit. We might rather think that if we have values for the weights, we also might have values with accompanying interpretation for the extrema. In the first case, these values are assumed to be probabilities. A value of 0.8 assigned to the pair (a, c) means that with a chance of 0.8, the pair will belong to the relation, or the arc will be present. We can also assume that it was the result of a pairwise matching experiment, where 80% of the subjects considered a to be in relation to c (for example, they preferred a over c if these would be the only options of a choice). Since then, 40% would have preferred c and $80 + 40 > 100$, there is also a number of indifferent choices.

Now, let us go for the greatest elements. For example, what is the probability then that a is the greatest element? It will be the greatest element if the following event happens: the pair (a, b) belongs to the relation, and the pair (a, c) also belongs to the relation. The former happens with a probability of 0.3 and the latter with a probability of 0.8. So the chance to have the greatest a in the relation sample is $0.3 \cdot 0.8 = 0.24$. In the same way, we find

a is greatest: $0.3 \cdot 0.8 = 0.24$
b is greatest: $0.6 \cdot 0.2 = 0.12$
c is greatest: $0.4 \cdot 0.5 = 0.20$

From this evaluation, it appears that a has the greatest chance to be a greatest element. But nevertheless, also b and c can be greatest with some probability that is not very different from that for a. In comparison, c is second-most likely and b least likely.

Now we consider maximal elements. The issue here is a little bit more complicated, since we have to take the asymmetric part into account. It means we have to first decide by which chance a pair belongs to the asymmetric part, an event that depends on the occurrence of a different pair, its converse. Consider the pair (a, b). It belongs to a relation sample with probability 0.3, while the pair (b, a) belongs to the same sample with a probability of 0.6. Then, the event for (a, b) being an element of the asymmetric part happens whenever (a, b) belongs to the relation (with a chance of 0.3) and (b, a) not (with a chance of $1.0 - 0.6 = 0.4$). Thus, (a, b) belongs to $P(R)$ with a chance of $0.3 \cdot 0.4 = 0.12$. If (a, b) is in the asymmetric part, it also means that b is not maximal. There is another case that would exclude b from the maximum set: if (c, b) would belong to the asymmetric part. In the same manner, we can find that this happens with a probability of $0.5 \cdot (1.0 - 0.2) = 0.40$. The event we need to ensure that b is maximal is: (a, b) does not belong to the asymmetric part, and (c, b) does not belong to the asymmetric part. It happens with probability $(1.0 - 0.12)(1.0 - 0.4) = 0.528$. So there is about a 50% chance that b is among the maximal elements. In the same manner, we can compute the other probabilities and obtain

a is maximal: 0.5336
$b$is maximal: 0.528
c is maximal: 0.468

As for the chance to be greatest element, it is largest for a, but the following order is reversed: next is c and then b. Also here, the differences are not very strong.

In order to explain these rather small differences, one might have a closer look at Fig. 2.12 to note that there is a preference cycle $a \to c \to b \to a$ in the figure, if just following the arcs with greater values than their converse arcs. So as a crisp relation, e.g., thresholded at 0.5, we would neither have maximal nor greatest elements. So the introduction of values alleviates the cycle-problem, at least a little. We note that the cycle problem will survive if we choose all values to be equal.

We have just examined the case where the values are interpreted as probabilities. Now we consider a second case where these values are seen as degrees by which the pair belongs to the relation. This can express a degree of certainty about the true presence of the relativeness of some pair, the confidence, its relevance, etc. If we want to consider extremality here as well, we first have to do some homework. The question is how we commit operations with the degree values in accordance with logical reasoning. We have defined greatestness and maximality purely on a logical base. It goes like: a is the greatest element if (a, b) belongs to the relation as well as (a, c). So we used logical AND for two predicates. What we need before we can process the degree values is the reflection of logical operations (esp. AND, OR, and NOT) in this processing. No need to say that there are many ways to do such a mapping between logical operators and arithmetic operations of the values. We will take here a most simple example.

If we consider the degrees as fuzzy membership values, we can use the minimum operator for the logical AND, the maximum operator for the logical OR, and subtracting the membership value from 1 for the logical NOT. Actually, this is all what we need to find out about extreme elements of a such-valued relation.

Let us first consider the greatest element, based on AND-ing all degrees of some element to be in relation to any other. So we just have to take the minimum of all these degree values. In our example, when considering the degree of element a being greatest appears as the minimum of the degree of (a, b) and the degree of (a, c)—which is 0.3. For visualization, for some vertex take the smallest value among all out-bound arcs emanating from it. So it is 0.3 for a, 0.2 for b and 0.4 for c. Now the winner is c as the greatest element to the highest degree, followed by a and b.

Maximality is more intricate. Let us find it in a top-down manner:

- We need the degree of the coincidence: (b, a) is NOT in the asymmetric part AND (c, a) is NOT in the asymmetric part.
- The degree of (b, a) is NOT in the asymmetric part is 1 minus the degree of (b, a) being in the asymmetric part.
- For (b, a) being in the asymmetric part, we need (b, a) being in the relation AND (a, b) being NOT in the relation.
- For the former we have the degree 0.6, the latter is of degree $1.0 - 0.3 = 0.7$. The logical AND is reflected by the minimum of both degree values, which is 0.6. Thus, (b, a) is in the asymmetric part to the degree 0.6 and NOT in the asymmetric part to the degree 0.4.
- In the same way we find that (c, a) is not in the asymmetric part to degree 0.8.
- Since (b, a) is not in the asymmetric part to degree 0.4 and (c, a) to degree 0.8 it follows that a is maximal to degree 0.4, the minimum of 0.4 and 0.8.

We can do the same evaluation for b and c and find that b is maximal to degree 0.5 and c to degree 0.4 (same as a). By means of degree of maximality we find that b is of highest degree (but also here there is not a big difference to the other two).

A small remark: did our choice of degree 1 for the pair (a, a) really not influence the evaluation? We can see in the same way that the degree of (a, a) being in the asymmetric part is the minimum of the degree of (a, a) being in the relation (which is 1) and (a, a) being NOT in the relation, which is 0. Then, (a, a) does NOT belong to the asymmetric part with degree 1. Indeed, this value does not affect the minimum value with the other degree values.

This example shows two things:

- The specification of extrema depends strongly on the assigned processing of the values. A process that is consistent with the Boolean algebra is needed. If such a specification is not possible, then there is no way of specifying extreme elements.
- Valued relations (i.e., numerically valued) appear as another method to circumvent circles (so to say). If we feel able to introduce values for the relational pairs, there is the advantage that each element can become greatest or maximal to some degree, with some chance etc., depending on the specific way to interpret the values.

Chapter 3
Resource-Aware Intelligent Communication Systems: Case Studies

3.1 Introduction

The advancement of technology has been expedited with the advent of high-end computational devices and storage systems. In addition to higher computational avenues, modern-day technology resorts to a plethora of data-capturing and acquisition devices, including a wide range of sensors and actuators. As a result, the amount of data being generated and needed to be processed has increased manifold, calling for greater computational resources to undertake the processing tasks. This has not only led to a huge demand for resources but also increased energy usage. Thus, the need for resource-aware systems has assumed paramount importance to optimize resource usage under user-specified constraints. Such resource-aware systems are equipped with the ability to monitor resource usage and can manage the resources according to user-specified constraints [13] by paying attention to resources such as power, bandwidth, and memory, to name a few [18].

The obvious fallout of the present-day technological innovations has led to several stressful issues, which include environmental sustainability, optimization of energy usage, and reduction of computational time [1]. Moreover, with the exponential increase in the variety of execution environments and application contexts, developers and practitioners have resorted to repeated refactoring of modern software to align the applications with ever-changing non-functional requirements [18]. This has led to issues of scalability and reliability. A self-adaptive, scalable, resource-efficient software has been envisioned, which can reconfigure itself with minimal programmer-specified objectives and resource constraints of the execution environment [18].

This chapter illustrates several case studies for resource-aware intelligent communication systems, highlighting their essence and applications.

M. Köppen et al., *Resource-aware Intelligent Communication Systems*, Springer Tracts in Human-Centered Computing, https://doi.org/10.1007/978-3-032-06915-3_3

3.2 Case Studies

3.2.1 Resource-Aware Memory Assessment Model for Memory-Constrained Context-Aware Systems [2]

Context-aware computing has enabled the application of sensors, actuators, and other computing devices in smart spaces for data acquisition and rendering intelligent services to users. The working memory size often becomes a critical factor in developing these systems, especially in resource-constrained situations. A rule-based system has been evolved in [2] to efficiently calculate the working memory size in this regard. The authors focussed on the dynamic working memory for this purpose. A newly created context was added to the dynamic working memory by firing a rule instance received as a message from another device or agent. Three different algorithms, viz., Distinct Working Memory (DWM), Average of the Preference Sets (APS), and Smart Average of the Preference Sets (SAPS), have been proposed to address the memory requirement problem.

The DWM algorithm considers the number of unique consequences of the rules rather than every consequence occurrence while estimating the memory requirement.

The APS algorithm considers the average of user preferences to tune the resource-limited context-aware applications. Since this algorithm banks on preferences, the memory space requirements are saved appreciably. Firstly, this algorithm determines the average working memory size using the overall average of rule base sizes. Finally, it identifies distinctive consequences in the rule base corresponding to the preference sets with working memory (WM) requirements greater than the estimated average.

The SAPS algorithm also focuses on preference sets. However, it optimizes the memory computation by calculating the standard deviation of the available preference sets. For standard deviation values less than a predetermined threshold value of 2, the algorithm selects the preference set from the available sets with the most memory units. For larger values of standard deviation, SAPS operates similarly to APS.

Evaluation results on two different systems (one a Core I5 system and another Raspberry PI system) using 153 rules (obtained by combining three different rule sets [38] indicate that the DWM algorithm consumes more memory than APS and SAPS, thereby validating the effectiveness of the proposed rule-based systems.

3.2.2 Smart Parking Lighting Assistance System (SPLAS) Using the VEINS Vehicular Network Simulation Framework [21]

The Internet of Things (IoT) envisages the seamless connectivity of billions of IoT devices. Thus, a scalable IoT architecture is required to efficiently manage many nodes in a heterogeneous IoT architecture.

A Smart Parking Lighting Assistance System (SPLAS) architecture has been proposed in [21] using the VEINS vehicular network simulation framework. The proposed architecture is a novel Internet of Things (IoT) based smart parking solution that offers tailored visual guidance to each driver inside large parking areas using Variable Message Signs (VMS). The architecture enables drivers to identify empty parking spaces near their desired destination, minimally using self-organization, considering the current traffic conditions. The VEINS simulation framework used to model the individual SPLAS node deployment, communication, and vehicle mobility accommodates different scenarios and parking area layouts.

Since the proposed SPLAS architecture is aimed at offering visual guidance to drivers, several VMSs are deployed over the intersections in the parking area. The constituent SPLAS nodes are deployed only over the multiple-direction intersections to enable drivers to make route decisions. Furthermore, the desired destination is selected by the drivers using a ticket (with a compact tracking module) from a vending machine.

In order to provide tailored directions to each vehicle, Visible Light Beacons (VLBs) are transmitted from the SPLAS nodes in a defined radius around them in periodic intervals so that any vehicle entering a node VLB area would receive a beacon containing the node ID, thereby returning the vehicle position announcement back to the SPLAS node. In turn, the SPLAS node uses this information and assists the driver with a possible parking direction using VMS. The SPLAS node also estimates the traffic during the process by means of self-supervision.

3.2.3 *Energy-Efficient Allocation of Resources in Multi-hop Multimedia Routing [1]*

Device-to-device (D2D) communication systems mandate energy-efficient routing. Hence, faithful allocation of resources is required in D2D communication systems to increase battery lifetime and quality of services (QoS).

A novel multi-hop routing mechanism based on combinatorial optimization has been proposed in [1] to this effect. An infinite planar cellular network is used in this mechanism to provide two alternative transmission modes, viz., (i) cellular mode and (ii) D2D mode, for the user equipment (UE). The proximal UE directly communicates with D2D without any knowledge of the base station BS by means of a Smart Edge Device in the cellular mode of transmission. On the other hand, adjacent UE can be discovered through the Received Signal Strength (RSS) in the D2D mode.

Interference is reduced considerably, and resource efficiency is enhanced for proximal UE transmission in D2D by resorting to spatial reuse. The communication link reduces the interference under various transmission modes by employing orthogonal resources. Simulation of the proposed mechanism demonstrates reduced computational costs and improved energy efficiency, throughput rate, resource utility, and packet delivery ratio compared to the state-of-the-art techniques.

3.2.4 *Semantic-Aware Intelligent Network System for Data-Centric Adaptive Spectrum Management in Urban Wireless Networks [15]*

The evolution and adoption of connected devices have resulted in an explosion of data-intensive applications. This, in turn, has exacerbated the challenges of spectrum and resource allocation in urban wireless networks.

A semantic-aware prioritization methodology has been envisioned in [15] to evolve an intelligent data-centric framework for effective resource management. The proposed framework, also referred to as the Semantic-Aware Intelligent Network System (SA-INS), ensures optimal allocation of information by evaluating the contextual content of data. The proposed system also utilizes a Q-learning-based algorithm to address the specific requirements of semantic-aware communication systems. The resultant system is found to promote scalability and optimize resource allocation using the non-dominated sorting genetic algorithm II (NSGA-II) based multi-objective framework. In addition, 5G/6G network specifics are incorporated in conjunction with a water-filling power allocation algorithm, enhancing spectrum efficiency in the process. Virtual networks are also created to meet the requirements for specific data types, thereby affecting efficient quality-of-service (QoS) provisioning.

The proposed system has been found to outperform the state-of-the-art methods as far as spectrum efficiency, latency, and packet loss rates (PLR) are concerned.

3.2.5 *Machine Learning Assisted Risk-Aware Resource Allocation Strategy for Ultra-Reliable Low-Latency Communication Systems [4]*

A distributed risk-aware machine learning solution is proposed in [4] for the radio resource management problem often encountered in ultra-reliable low-latency communications (URLLC). Proper and effective scheduling of identified shared and dedicated resources is required to efficiently manage network resources for serving scheduled and non-scheduled URLLC traffic. The proposed solution is centered on determining the rank of the underlying users based on the delay budget and the expected level of received interference over the identified shared resources. The machine learning-based solution to the radio resource management problem entails (i) radio resource provisioning (RRP), (ii) radio resource scheduling (RRS), and (iii) radio resource utilizing (RRU).

The partial observability of the base stations (BSs) in cellular networks is taken care of by distributing the learning task between BSs and the radio access network (RAN) control center. Since the BSs only deal with local information, they are less effective when it comes to resource provisioning for mobile URLLC users. The proposed distributed learning framework incorporates effective resource scheduling policies in these circumstances. This distributed framework also enables the BSs to monitor instant changes in the channel quality of connected users and schedule them based on the remaining delay budgets of the users.

Experimental results demonstrate the superior performance of the proposed solution compared to the conventional approaches.

3.2.6 An Intelligent QoS-Aware Resource Management Approach for Cloud Computing [10]

Effective management of resources in cloud computing is a challenging thoroughfare, thanks to the uncertainty and dispersion of resources.

An intelligent approach for quality-of-service (QoS) based autonomic resource management is proposed in [10] to ensure reliable, secure, and cost-efficient cloud services through an automatic management of the underlying resources. The proposed approach, referred to as CHOPPER (Configuring, Healing, Optimizing and Protecting Policy for Efficient Resource Management), self-configures applications and resources, self-heals sudden failures, self-protects against security attacks, and self-optimizes resource allocation for maximum utilization.

CHOPPER manages the QoS requirements of cloud users automatically to reduce human intervention. The provisioned cloud resources are scheduled efficiently to offer seamless cloud service delivery by avoiding service level agreement (SLA) violations. CHOPPER also optimizes execution cost, time, energy efficiency, and resource contention in the process. Workloads are classified into (a) normal (relaxed deadline), (b) good (average deadline) and (c) critical (urgent deadline) instead of employing the first cum first serve workload execution model.

Performance evaluation of the proposed approach in a real cloud The environment reveals the superiority of the proposed approach over the state-of-the-art approaches as far as cost, execution time, SLA violation, resource contention, and security are concerned.

3.2.7 A Deep Learning Framework for Situation-Aware Resource Allocation for Multidimensional Intelligent Multiple Access [23]

Proper utilization of resources is essential in communication systems with diverse requirements. Hence, a situation-aware intelligent strategy for the utilization of resources is required in multidimensional communication systems.

The authors presented a deep learning-based framework to enable intelligent resource allocation in beyond 5G and 6G wireless networks in [23]. The proposed multidimensional intelligent multiple access (MD-IMA) scheme is aimed at addressing the real-time and diverse quality-of-service (QoS) requirements of these networks through the utilization of the available radio resources in heterogeneous domains.

Long short-term memory (LSTM) neural network and deep reinforcement learning (DRL) frameworks form the essence of the proposed intelligent system. The LSTM neural network predicts the long-term network dynamics and inference changes in the QoS requirements. A deterministic policy gradient (DDPG) algorithm monitors the fluctuations of the network situation and optimizes the multidimensional radio resource allocation.

The proposed deep learning framework is found to outperform state-of-the-art algorithms on real-world cellular traffic datasets.

3.2.8 A Digital Twin Architecture for QoS-Aware Intelligent Industrial Communication [5]

Safety and security of operations and communications are key requirements in modern industrial networking environments. IEC 62443 zones and conduits mandate clear separation of machines for safety and inter-machine communication security requirements. Conventional models are often prone to human-centric errors.

An application of a digital twin approach to the zones and conduits is presented in [5] to facilitate the definition and management of inter-machine security requirements. This enables industrial technicians to ascertain that the communication between two zones always flows through a ciphered conduit with a given algorithm and key length. Moreover, an intelligent reasoner can be implemented to monitor the current state of the assets and networks through the digital twin. This reasoner helps in reconfiguring the environment in the event of non-satisfaction of the desired requirements, thereby enforcing specific requirements and fulfilling a proper trade-off between security and performance.

Evaluation of the proposed prototype demonstrates the efficiency of the proposed solution under constrained industrial environments.

3.2.9 A Multi-objective Framework for Traffic-Awareness Dynamic Resource Allocation in Multi-beam Mobile Satellite Communication Systems [12]

Mobile satellite communication systems mostly operate at the L or S band with multiple beams reusing the limited spectrum efficiently. This frequency reuse mechanism gives rise to co-channel interference issues, which result in the coupling of the bandwidth allocation, thereby posing a serious challenge to optimize the resource allocation mechanism.

The authors presented a novel multi-objective framework based on the newly proposed traffic-aware dynamic resource allocation (TADRA) algorithm [12] to optimize resource allocations dynamically, thereby ensuring higher quality-of-service

(QoS) for multiple services of different priorities. The Pareto-front optimal solutions are derived with respect to the co-channel interference and onboard transmit power. Experimental results indicate that a well-balanced trade-off is achieved between high-priority call completion ratio and video and data throughput in medium and low priorities. Moreover, the proposed multi-objective framework attains the Pareto-front solutions faster.

3.3 Conclusion

This chapter puts forward several indigenous resource-aware intelligent approaches for different types of communication systems. The representative case studies span across diversified domains of communication systems, ranging from memory-constrained systems to mobile satellite communication systems. The efficacy of the proposed cases is found to be superlative when compared to the conventional state-of-the-art systems. Although a few case studies have been reported in this chapter, it is evident that the resource-awareness in the existing communication systems can be enhanced using intelligent tools and techniques.

References

1. Al-Turjman F, Deebak BD, Mostarda L (2019) Energy aware resource allocation in multi-hop multimedia routing via the smart edge device. IEEE Access 7:151203–151214
2. Ali M, Arshad M, Uddin I, Ali G, Asim M, Elaffendi M (2024) A resource aware memory requirement calculation model for memory constrained context-aware systems. IEEE Access 12:19320–19329
3. Axelrod R (1990) The evolution of co-operation. Penguin Books
4. Azari A, Ozger M, Cavdar C (2019) Risk-aware resource allocation for urllc: challenges and strategies with machine learning. IEEE Commun Mag 57(3):42–48
5. Bellavista P, Giannelli C, Mamei M, Mendula M, Picone M (2022) Digital twin oriented architecture for secure and qos aware intelligent communications in industrial environments. Pervasive Mob Comput 85(101646):1–17
6. Bertsekas D, Gallager R (1992) Data networks, 2nd edn. Prentice Hall
7. Bouyssou D (1992) Ranking methods based on valued preference relations: a characterization of the net flow method. Eur J Oper Res 60(1):61–67
8. Defays D (1979) Tree representations of ternary relations. J Math Psychol 19(2):208–218
9. Frank RH, Cook PJ (2010) The winner-take-all society. Virgin Books
10. Gill SS, Chana I, Singh M, Buyya R (2018) Chopper: an intelligent qos-aware autonomic resource management approach for cloud computing. Clust Comput 21:1203–1241
11. González-Pachón J, Romero C (2008) A method for obtaining transitive approximations of a binary relation. Ann Oper Res 163(1):197–208
12. He YZ, Jia YZ, Zhong XD (2017) A traffic-awareness dynamic resource allocation scheme based on multi-objective optimization in multi-beam mobile satellite communication systems. Int J Distrib Sens Netw 13(8):1–14
13. Huang W-C, Knottenbelt WJ (2014) Low-overhead development of scalable resource-efficient software systems. In: Handbook of research on emerging advancements and technologies in software engineering, pp 81–105. IGI Global, USA
14. Isard M, Prabhakaran V, Currey J, Wieder U, Talwar K, Goldberg A (2009) Quincy: fair scheduling for distributed computing clusters. In: Proceedings of the ACM SIGOPS 22nd symposium on operating systems principles, pp 261–276
15. Islam A, Chang K (2020) Pushing the boundaries of urban wireless networks: a semantic-aware intelligent network system (sa-ins) for data-centric adaptive spectrum management. IEEE Open J Commun Soc 1–24
16. Jain RK, Chiu D-MW, Hawe WR et al (1984) A quantitative measure of fairness and discrimination. Eastern Research Laboratory, Digital Equipment Corporation, Hudson, MA, vol 21, p 1

M. Köppen et al., *Resource-aware Intelligent Communication Systems*, Springer Tracts in Human-Centered Computing, https://doi.org/10.1007/978-3-032-06915-3

17. Jiang J, Sekar V, Zhang H (2012) Improving fairness, efficiency, and stability in http-based adaptive video streaming with festive. In: Proceedings of the 8th international conference on emerging networking experiments and technologies, pp 97–108
18. Jiang W, Jin H, Yu C, Liu C (2010) Introduction and overview of wireless sensor networks. In: Handbook of research on developments and trends in wireless sensor networks: from principle to practice, pp 1–19. IGI Global, USA
19. Kalai E, Smorodinsky M (1975) Other solutions to nash's bargaining problem. Econometr: J Econometr Soc 513–518
20. Karp RM (1972) Reducibility among combinatorial problems. Springer
21. Kekempanos A (2023) Smart parking lighting assistance system (SPLAS). PhD thesis, Eindhoven University of Technology
22. Lan T, Kao D, Chiang M, Sabharwal A (2009) An axiomatic theory of fairness in network resource allocation. arXiv:0906.0557
23. Liu Y, Wang X, Mei J, Boudreau G, Abou-Zeid H, Sediq AB (2021) Situation-aware resource allocation for multi-dimensional intelligent multiple access: a proactive deep learning framework. IEEE J Sel Areas Commun 39(1):116–130
24. Marshall AW, Olkin I, Arnold BC (2011) Inequalities: theory of majorization and its applications, 2nd edn. Springer
25. Maurer SB (1980) The king chicken theorems. Math Mag 53(2):67–80
26. Nash JF (1950) Equilibrium points in n-person games. Proc Natl Acad Sci 36(1):48–49
27. Nishimura H (2012) The transitive core of preference relations. Available at SSRN 2059254
28. Novák V, Novotnỳ M (1989) Transitive ternary relations and quasiorderings. Archivum Mathematicum 25(1):5–12
29. Novák V, Novotnỳ M (1992) Binary and ternary relations. Math Bohem 117(3):283–292
30. Saaty TL (1980) The analytic hierarchy process: planning, priority setting, resource allocation. McGraw-Hill
31. Saaty TL, Forman E (1993) The hierarchon. A Dictionary of Hierarchies. RWS Publications, Pittsburgh
32. Schäfer PM (2012) Betweenness relations. PhD thesis, Universität Ulm. Fakultät für Mathematik und Wirtschaftswissenschaften
33. Schmidt G (2011) Relational mathematics. Cambridge University Press
34. Sharir M (1981) A strong-connectivity algorithm and its applications in data flow analysis. Comput & Math Appl 7(1):67–72
35. Suzumura K (1983) Rational choice, collective decisions, and social welfare. Cambridge University Press
36. Tarjan R (1972) Depth-first search and linear graph algorithms. SIAM J Comput 1(2):146–160
37. Thurstone LL (1927) A law of comparative judgment. Psychol Rev 34(4):273
38. Uddin I (2019) A rule-based framework for developing context-aware systems for smart spaces. PhD thesis, University of Nottingham
39. Wang Y, Ma W, Zhang M, Liu Y, Ma S (2023) A survey on the fairness of recommender systems. ACM Trans Inf Syst 41(3):1–43
40. Young HP (1994) Equity: in theory and practice. Princeton University Press
41. Zitzler E, Laumanns M, Thiele L (2002) Spea2: improving the strength pareto evolutionary algorithm for multiobjective optimization. In: Evolutionary methods for design, optimisation, and control, CIMNE, Barcelona, Spain, pp 95–100

Index

M. Köppen et al., *Resource-aware Intelligent Communication Systems*, Springer Tracts in Human-Centered Computing, https://doi.org/10.1007/978-3-032-06915-3

V

W

MIX
Papier aus verantwortungsvollen Quellen
Paper from responsible sources
FSC® C105338

If you have any concerns about our products,
you can contact us on
ProductSafety@springernature.com

In case Publisher is established outside the EU,
the EU authorized representative is:
Springer Nature Customer Service Center GmbH
Europaplatz 3, 69115 Heidelberg, Germany

Printed by Libri Plureos GmbH
in Hamburg, Germany